BIOMEDICAL GRADUATE SCHOOL: A PLANNING GUIDE TO THE ADMISSIONS PROCESS

David J. McKean
Professor Emeritus of Immunology
Mayo Clinic

Ted R. Johnson
Director of Biomedical Studies
St. Olaf College

JONES AND BARTLETT PUBLISHERS
Sudbury, Massachusetts
BOSTON   TORONTO   LONDON   SINGAPORE

World Headquarters

Jones and Bartlett
  Publishers
40 Tall Pine Drive
Sudbury, MA 01776
978-443-5000
info@jbpub.com
www.jbpub.com

Jones and Bartlett
  Publishers Canada
6339 Ormindale Way
Mississauga, Ontario L5V 1J2
Canada

Jones and Bartlett
  Publishers International
Barb House, Barb Mews
London W6 7PA
United Kingdom

Jones and Bartlett's books and products are available through most bookstores and online booksellers. To contact Jones and Bartlett Publishers directly, call 800-832-0034, fax 978-443-8000, or visit our website, www.jbpub.com.

Substantial discounts on bulk quantities of Jones and Bartlett's publications are available to corporations, professional associations, and other qualified organizations. For details and specific discount information, contact the special sales department at Jones and Bartlett via the above contact information or send an email to specialsales@jbpub.com.

Production Credits
Publisher, Higher Education: Cathleen Sether
Acquisitions Editor: Molly Steinbach
Managing Editor: Dean W. DeChambeau
Editorial Assistant: Caroline Perry
Production Director: Amy Rose
Production Assistant: Laura Almozara
V.P., Manufacturing and Inventory Control: Therese Connell
Composition: Auburn Associates, Inc.
Cover Design: Scott Moden
Cover Image: © Cre8tive Images/ShutterStock, Inc.
Printing and Binding: Malloy, Inc.
Cover Printing: Malloy, Inc.

Library of Congress Cataloging-in-Publication Data

McKean, David J.
  Biomedical Graduate School : a planning guide to the admissions process /
David J. McKean and Ted R. Johnson
    p. cm.
  Includes bibliographical references.
  ISBN 978-0-7637-6000-7
  1. Medical colleges—Admission. 2. Medical sciences—Study and teaching
(Graduate) 3. Doctor of philosophy degree. 4. Graduate students. I. Johnson,
Ted R., 1946- II. Title.
  R838.4.M45 2009
  610.71′173—dc22

                          2008038416

6048
Printed in the United States of America
13   12   11   10   09   10   9   8   7   6   5   4   3   2   1

DEDICATION

We dedicate this book to the undergraduate and graduate school faculty who invested their time and energy to train us to be educators and biomedical research investigators during our professional education and careers. We have had the opportunity to work with many creative, highly skilled, and caring individuals. We greatly appreciate the many contributions these individuals have made to our professional and personal lives.

ABOUT THE AUTHORS

David J. McKean, PhD, is Professor Emeritus of Immunology at the Mayo Clinic in Rochester, Minnesota. During his 30 years as a National Institutes of Health (NIH)–funded investigator on the research faculty of Mayo Clinic, he published 134 peer-reviewed research publications and was actively involved in PhD and MD-PhD education programs both as a student mentor and administrator. His academic administrative responsibilities included Dean of Mayo Graduate School, Chair of the Mayo Immunology Department, and Admissions Committee member of both the Mayo Graduate School PhD program and Mayo MD-PhD program. He also participated as a reviewer of graduate and postgraduate fellowship applications and faculty research grants sponsored by the Howard Hughes Foundation, American Heart Association, American Cancer Society, and National Institutes of Health.

Ted R. Johnson, PhD, is a Professor of Biology and the Director of Biomedical Studies at St. Olaf College in Northfield, Minnesota. Ted has been involved in undergraduate education and advising students for 35 years. He is the Chair of the Health Professions Committee, which prepares committee evaluations for students pursuing careers in medicine, dentistry, and other vocations in the biomedical arena. He has served as Chair of the Biology Department and was the Associate Dean of Students. He has participated as a reviewer of graduate fellowship applications sponsored by the Howard Hughes Foundation and undergraduate research grant applications for the National Science Foundation.

BRIEF
CONTENTS

CONTENTS

List of Feature Boxes

PREFACE

This book is based on the professional experience of two science professors who have 65 years of combined experience in mentoring undergraduate students and recruiting students to PhD and MD-PhD programs.

Many undergraduate students who have an interest in science careers do not get appropriate information on how to identify potential career options or to maximize their academic and research opportunities during different periods in their academic development. Undergraduate students need to carefully plan their selection of academic classes and their extracurricular activities to maximally explore the wide range of potential career options in science. If they choose a professional career in science, they need to participate in an advanced degree program after completing their undergraduate education. The quality of the advanced degree program they select will likely have a significant impact on their future career. Because many applicants are inadequately prepared for the graduate school application and interview process, their applications to advanced degree programs do not represent the students' academic credentials. Consequently, students may either end up not being accepted by competitive graduate programs or matriculating into less-competitive programs. These shortcomings can be significantly improved as a result of step-by-step guidance provided in this book.

We identify what college students need to do throughout their undergraduate education to become competitive for and be accepted into a biological or biomedical PhD or MD-PhD program. We describe the application and interview process in detail and coach students on how to maximize their academic credentials. Finally, we provide strategies to help applicants select an education program to enhance their career options or, if they are not accepted to their selected schools, to improve their credentials for subsequent rounds of application. Although the text is primarily directed at students, it is also useful to undergraduate faculty in mentoring and preparing letters of recommendation for their students.

The book begins by helping students identify potential career options that align with their intellectual interests, abilities, and personal quality-of-life issues. It further addresses career options by evaluating the spectrum of career opportunities that are available to PhD and MD-PhD graduates in the biological and biomedical sciences. It outlines the details of typical PhD and MD-PhD education programs so that students understand the breadth and depth of the academic programs they will undertake to complete these advanced degrees. It provides valuable information to help undergraduates select their education curriculum, extracurricular activities, and the research experiences they will need to attain. It provides goals and expectations to students at different stages of their undergraduate education, to enable them to benchmark their academic progress.

Deciding where to apply for a PhD or MD-PhD degree program should be a rational process. To accomplish this, we make suggestions to help align a student's academic, clinical, and research interests with the academic curriculum, clinical strengths, research interests, and productivity of various advanced degree educational programs. In addition to providing information on the criteria that should be used to select schools where the student will apply, we explain the preparation of PhD and MD-PhD applications in great detail. We describe each step of the graduate education application process, from the perspective of both the student and the faculty member(s) who will review the application. Suggestions about what to write and what not to write help the student prepare competitive applications, and a separate section discusses what faculty should and should not include in letters of recommendation. We also guide students through one of the most neglected aspects of the application process—the interview. Our step-by-step guides help the student prepare before making a campus visit and provide strategies for handling difficult situations during the interview day. Finally, we help students prioritize the issues, to enable them to select an education program that meets their career goals in science. Although the book is directed primarily at students seeking advanced degrees in the biological or biomedical sciences, the information on selecting a graduate school, preparing applications, and basic interviewing skills will be applicable to students applying to any graduate school or medical school.

CHAPTER 1

There Are So Many Career Pathways in Science: How Do You Identify Your Options?

Selecting a career path can be a challenging endeavor for an undergraduate student. When you start considering your potential career path, you may be overwhelmed with all the choices. What career matches your interests and your abilities? How do you decide what to major in and what courses to take? What experiences outside of the classroom will be beneficial to you as you develop career choices? You want to identify a career that aligns with your intellectual interests and abilities as well as addresses the important quality of life issues that you develop during your academic training. If you are a first-year student in college, you have limited experiences to help you identify your intellectual interests, understand your abilities, or become exposed to important quality of life issues. We focus our initial discussion on these three issues that are critical to identifying your potential career options. Although we focus our discussion on identifying potential careers in science, most issues we consider are also applicable to many other non-science career pathways.

■ IDENTIFYING QUALITY-OF-LIFE ISSUES THAT ARE IMPORTANT TO YOU

Let's first discuss how your educational experiences will help you understand important quality-of-life issues—what are the important components in your life that may make you unique from others? During your first year of college, most of your education will likely be in introductory classes in the arts, humanities, and social and/or natural sciences. These classes are designed to enable you to acquire oral and written communication skills and to expose you to a breadth of areas that contribute to your life experiences, critical thinking skills, intellectual development, and becoming an independent learner. Although your introductory classes may or may not expose you to new potential career options, the classes should help you develop important values and intellectual passions that contribute to making you a unique individual. Start the process of self-discovery with an open mind. You may view these introductory courses as bottlenecks that you must pass through before you can take high-level classes that teach professional career skills. However, technical skills are of limited value unless you also have developed creative problem-solving skills and a strong set of values to help you sort through the contradictory issues you may encounter in your personal and professional life. It is unlikely that you will have another opportunity in your life to encounter the breadth of issues that are addressed in traditional liberal arts classes. These educational experiences can contribute significantly to your personal intellectual development.

Although classroom courses are the academic base of your undergraduate education, participation in internships, seminars, volunteer activities, service opportunities, socializing, and intramural activities can significantly contribute to the breadth of your education. As you experience these activities, you enhance your interpersonal communication skills and, importantly, you learn about your personal interests, strengths, and weaknesses. As you progress through your college years, pause to reflect and consider what is important to you and why it is important.

■ HOW DO YOU IDENTIFY AREAS OF SCIENCE THAT ARE OF INTEREST TO YOU?

The curriculum at most undergraduate colleges starts out by providing classes in many different areas of the arts and sciences. At this point in your education you may not have a clue about the type of career you want to pursue. By the end of the first or second year of college, you should have sufficient educational experience to select your undergraduate course major and begin to narrow your education focus. What areas of science do you enjoy and would you like to pursue? Identifying potential career options is easier to approach once you have identified areas in science that interest you. Later in this chapter we discuss strategies to help you explore various career options. However, early in your undergraduate education you should keep many career options open while you decide whether or not you will be a science major. Subsequently, you should identify the broad area of science (biology, chemistry, engineering, etc.) where you will focus your academic studies for the next several years.

■ HOW DO YOU IDENTIFY YOUR INTELLECTUAL INTERESTS?

As you participate in the formal and informal components of your undergraduate education, you become exposed to areas of science that are more interesting to you than others. You may have a class, a seminar, or a volunteer activity that introduces you to an aspect of science that particularly stimulates your imagination. What interests you? Your faculty advisor is a valuable source of information to help you learn more about specific areas of science in your major. You will benefit from your advisor understanding your interests and abilities as well as using your advisor's experience and knowledge to efficiently increase your understanding of specific areas of science.

Many areas of interest develop as you progress in your academic studies and specific topics may attract your attention. One approach to explore a topic you are interested in is to do a computer search to learn of the potential medical or commercial applications. You also may want to identify researchers who are studying the topic and the questions they are asking. You may be surprised to learn about the different approaches investigators may be using to understand their projects. Most research institution websites have faculty research descriptions, and these research descriptions can be identified from key word analyses using computer search engines. As you gain experience as an undergraduate, you will likely identify multiple areas that potentially interest you. At this time in your education, your focus should be to explore different interests rather than focus on a narrow area of science. Perhaps you believe there are too many interesting areas and you are trying to determine which are the most interesting. To help you sort through this issue, talk to your advisor or to other faculty members. You may want to pursue some of these areas of interest by taking a class in the subject or by talking with individuals who are involved in the research that has caught your interest.

■ HOW TO EVALUATE YOUR PERSONAL ACADEMIC STRENGTHS AND WEAKNESSES

To take maximum advantage of the personal growth opportunities during the formal and informal components of your undergraduate education, you need to practice introspection. What are the academic activities you enjoy? Obtaining enjoyment from specific aspects of your educational experience usually occurs in areas in which you do well. Although you may have to work hard to accomplish activities that provide you with the most satisfaction, it is likely your innate and acquired abilities significantly contribute to being able to accomplish these activities. In contrast, most of us do not enjoy doing things that are difficult for us to accomplish. Critically evaluate how you interact with others and how you respond during the difficult times in your life. Take a few minutes at the end of every week to evaluate what you have done during the past 7 days, which activities or interactions gave you the most satisfaction, and consider what you have learned about yourself as a consequence of dealing with the positive and negative aspects of your recent activities. This self-analysis also can help you avoid repeating mistakes that you may have made. You may find it useful to keep a personal journal that records these introspective analyses. By rereading your journal writing from months past, you should be able to better understand how your strengths and weaknesses evolve as you acquire life experiences.

■ FIRST STEPS TO IDENTIFYING A POTENTIAL CAREER FOCUS

As your undergraduate education develops, you need to transition from identifying areas of science that interest you to exploring career options in areas that not only interest you but that fit your unique talents, abilities, and goals in life. A variety of resources and activities are available to help you align your interests, intellectual strengths, and personal goals to develop career options. In addition to a wealth of resources in college career centers, there is a wide range of useful information on the Internet. One of your initial objectives should be to discern how the characteristics and requirements of various career options potentially fit the science interests that you have identified from your undergraduate education.

Career information from professionals

A variety of approaches can be used to obtain information on careers. You may be able to take advantage of information sessions conducted by graduate school, medical school, or MD-PhD program directors held at your institution or at an open house at the graduate school. Although we discuss advanced degree programs in Chapter 2, program directors can provide you with insight concerning what scientists do in their careers and how they arrived in their present career position. These individuals can describe not only the structure of their advanced degree education programs and how their students develop their professional skills, but they should be able to explain to you the different career paths taken by their students. You also may want to ask about what they consider to be the most positive and most negative aspects of specific careers in science. As we explain in Chapter 2, research and teaching are not the only career options for graduates of advanced degree programs. At this point you should be primarily interested in comparing distinct career options rather than identifying a specific area of science to focus your interests. From a practical perspective, professors who will later be writing letters of recommendation for you need evidence that you are interested in a career in the biomedical sciences and can demonstrate a well-developed path toward that career. It is difficult for faculty to address your motivation for an advanced degree program in a letter of recommendation if you do not attend research seminars or are not seen by faculty as a student interested in a professional career in science.

If science research is one of the potential career options you have identified, attend departmental seminars conducted by professors who have research laboratories at your institution, other academic institutions, or industry. Does their research interest you? Talk with the speakers after their seminar to determine if their career path is similar to what you might be interested in once you obtain your undergraduate degree. Explore! Another invaluable source of information regarding matching a career with your interests is to conduct a series of informational interviews with professionals in your areas of interest. Try to meet with professionals in a variety of careers, such as teaching, academic research, industry, medical practice, regulatory agencies, and clinical testing laboratories. Your science faculty and the staff of your college career center should be able to help you arrange these interviews and help you develop a series of questions to ask in the interview. An informational interview gives you the opportunity to ask questions and gather information from professionals in different career areas. Ideally, the interview is an opportunity to visit them at their place of work and get a feel for their work environment, job responsibilities, team interactions, and career path.

Internships enable you to directly participate in career activities

Once you identify careers that potentially interest you, interact with your career center and your faculty advisor to set up multiple internships where you can shadow professionals for an extended period of time. The internship may develop into a research experience in the future. (Laboratory research experiences are discussed in Chapter 4.) Internships could be with research investigators at your institution, alumni of your institution, or other professionals identified by your career center, advisors, or family friends. Before you begin the internship, identify specific objectives of the internship with the help of your faculty advisor and develop a schedule of when you will participate in the internship. At the end of the internship, meet with your advisor to discuss how you met the specific objectives that were established at the start of the internship. Keep a journal as you progress through the internship. Also identify what you learned from the experience (the good, bad, and ugly) and how the internship has affected your perspective of potential career options. Although you may believe it is beneficial to identify just the positive aspects of the internship, you may learn just as much about your potential career decisions from identifying the aspects of your internship that you did not like.

■ ARE YOU INTERESTED IN A CAREER IN THE BIOLOGICAL AND BIOMEDICAL SCIENCES?

In this book we discuss in considerable detail education programs and career pathways that focus on the biological and biomedical sciences. The term *biomedical* encompasses a very large area of science.

As the word indicates, it includes both biology and medicine. Biomedical science include such diverse areas as using computer modeling to design new drugs that treat disease, using engineering approaches to design noninvasive imaging techniques, using field biology techniques in an ecosystem to identify potentially useful pharmacological components, or characterizing the genes that regulate your body's immune response to infection. Studies in biological and biomedical science may be theoretical or have practical applications that affect the environment or can be used to characterize the molecular basis of disease.

Do some of these broad areas of science interest you? Students who thrive in science careers share certain traits. They enjoy the intellectual challenge of solving problems and are persistent when faced with obstacles. They are independent, creative, have a passion for science, perform well in their undergraduate classes, and want to contribute to society by expanding existing scientific knowledge and educating others. Science excites these students, and they are eager to explore new approaches to solving incompletely understood issues in science.

■ SUMMARY

1. Your introductory undergraduate classes are designed to enable you to acquire oral and written communication skills and to expose you to a breadth of areas that contribute to your life experiences, critical thinking skills, intellectual development, and becoming an independent learner.

2. By the end of the first or second year of college, you should have sufficient educational experience to select your course major and begin to narrow your education focus.

3. As you participate in the formal and informal components of your undergraduate education, you become exposed to areas of science that are more interesting than others. Take time to explore areas that interest you.

4. To maximally take advantage of the personal growth opportunities during the formal and informal components of your undergraduate education, you need to be introspective.

5. Start early in your undergraduate education to explore potential career options and reflect on how these options fit your interests as well as your academic ability.

6. Interview science professionals to explore the details of their activities and responsibilities.

7. Participate in internships that will enable you to directly explore a career option.

8. Attend research seminars at your institution to identify applications of the areas of science you have found interesting in your undergraduate classes.

9. Demonstrate to your faculty that you have an interest in a research career by participating in career activities.

CHAPTER

2

How Much Education Do You Need?

You will spend at least 4 years completing the requirements of your undergraduate bachelor's degree. During this time you will consider what career may be appropriate for you. As an undergraduate you have the opportunity to explore many different career options. If you are in your first year of college, you are just beginning this process. If you are a college sophomore or junior, your classes, seminars, and discussions with faculty should have helped you identify some areas of science that you enjoy. The next step is to evaluate the potential career opportunities that may be related to these areas of science. We discuss approaches that you can use to identify potential career options during your undergraduate education. Our discussion focuses on how to evaluate different areas of science where you may want to continue during your graduate education. However, you can use the same approaches if you are trying to identify potential careers that do not require an advanced degree. As part of the evaluation of your career options, you must consider the amount of education needed to effectively compete for different jobs.

■ CAREER OPTIONS FOR INDIVIDUALS WITH DIFFERENT LEVELS OF EDUCATION

Once you graduate with your bachelor's degree, you must decide if you will seek employment immediately or continue your education in an advanced degree program. There are a wide variety of career options for individuals with a bachelor's degree in science. For example, you could work in a research laboratory as a technician, test samples in a clinical laboratory, teach middle school or high school science classes, work as a field biologist, or become a sales representative for a drug company. In general, these are entry level positions with relatively low salaries, and they may have limited opportunities for career growth and intellectual development. A master's degree in science may provide a higher salary range, but your long-term career options and salary range may not be greatly improved as compared with having a bachelor's degree. A master's degree in science is not a prerequisite to enter a PhD or MD-PhD degree program. Most students go directly from completing their bachelor's degree to initiating a PhD or MD-PhD degree program. In science, having a PhD or a MD-PhD degree opens up many career options for you that are difficult to attain with either a bachelor's or master's degree. In general, the higher the level of education and training that you attain, the more your career opportunities, salary, and professional independence are maximized. The PhD or MD-PhD advanced degree programs prepare you for a career that will challenge your intellect, contribute to improving the health of others, and provide a higher level of potential economic rewards. An intellectually challenging career often leads to happiness and satisfaction with your career choice.

It is not unusual for college graduates to become employed right out of college because they have identified a career they enjoy or they need to earn money to pay their college debts or pay for their current lifestyle. Others may become employed because they are not certain what they want to do and may want to try several jobs to gain experience to help them identify a career where they will do well and that satisfies their ambitions. If your college degree has prepared you for a job that you find satisfying, you are a very fortunate individual. However, if you decide several years after obtaining your undergraduate degree that you want a position that requires additional education, you will have multiple options to return to school either part time or full time. You could find that a 1- or 2-year program of specialized training, such as in the health sciences, will provide you with a job that will likely continue to be in demand and will contribute to society. The Mayo Clinic School of Health Sciences website is a good source of information on potential jobs in the health sciences (http://mayoweb.mayo.edu/mshs/).

Web Address Information

We provide a large number of Internet references in this book because, except for some academic sites, most contain the most current information. However, the website addresses of these sites may change with time. If the website you select no longer exists, try going to the stem of the site address and search for the desired topic. Thus for http://mayoweb.mayo.edu/mshs/, you could enter http://mayoweb.mayo.edu and, once on the site, search for "health sciences."

Alternatively, you may determine that you want a science career with more options, intellectual contributions, independence, responsibility, and financial opportunities. For many careers as a science professional, you need to obtain a PhD, an MD, or an MD-PhD degree (discussed later). These degree programs provide advanced education in a defined area of biological or biomedical research (PhD) or clinical medicine (MD). There are many career options for students with graduate degrees in the biological and biomedical sciences. These options include a wide array of disciplines such as biology, chemistry, biochemistry, physics, engineering, and computer science as well as the opportunity to become involved in interdisciplinary studies. Students who graduate from PhD graduate training programs in science have a wide range of job opportunities after they complete their training. There are traditional career options in an academic higher education institution as a laboratory research investigator, as an educator of undergraduate or graduate students, or some combination of these two options. In addition, there are many additional opportunities in clinical testing laboratories, government research laboratories, regulatory agency research laboratories, and oversight of product development and production in industry.

■ COMBINED DEGREE PROGRAMS

Physician scientists with an MD-PhD degree have many career options that may combine academic research, clinical practice and teaching, clinical research focused on studies with patients, or translational studies that combine laboratory and clinical research or patient-oriented research in industry. In addition to the traditional PhD or MD degree training programs, combined degree programs provide specialized training in an additional degree program that integrates biology or medicine with another career focus. There are a wide range of possible specialized training options that include clinical research, law, public health, health policy, informatics, divinity, epidemiology, translational research, and business (http://services. aamc.org/currdir/section3/degree2.cfm). Some of these combined degree programs include MD-PhD, MD or PhD plus Juris Doctor (J.D., law), Master of Public Health (MPH), Master of Science in Public Health (MSPH), Master of Health Informatics (MHI), Master of Divinity (MDiv), Master of Arts (MA) in Health Policy, Master of Clinical Epidemiology (MCE), Master of Translational Research (MTR), or Master of Business Administration (MBA).

As previously described, in addition to a PhD or MD degree, there are multiple degree options that can be taken as separate degrees or in combined degree programs. The combined degree programs are usually associated with medical schools (MD plus a second degree). Both the combined MD and master's degree programs can usually be completed in 5 years. Information on these programs can be found on the websites of medical schools that have these programs. A complete list of such medical schools is found on the AAMC directory website (http://services.aamc.org/currdir/ section3/start.cfm). If you are considering a PhD plus a second degree, you have a more limited number of schools from which to choose. You may have more academic options if you take the two degrees separately. If you are interested in a career that involves biological or biomedical research, you should explore what is involved in a PhD or a MD-PhD program. Such degree programs do not exclude a focus in additional career areas such as the master's degree programs described above. As your horizons expand with additional education, you may identify new interests that align with your education and career goals. Before deciding whether a PhD or PhD-MD program is appropriate for you, you need to understand what is involved in completing the degree requirements for a PhD or MD-PhD. We initially discuss the nature of the education program associated with completing the requirements for a PhD graduate degree. We subsequently consider how the PhD program differs philosophically and structurally from the MD-PhD program. Different career opportunities for individuals with various degrees and their potential salary considerations are also discussed.

■ HOW MUCH TIME WILL YOU NEED TO COMMIT TO ADVANCED DEGREE PROGRAMS?

The PhD graduate school program is a 4- to 7-year investment that prepares a student for a career as an independent research scientist. There are multiple career options available to individuals with PhD degrees in the biological or biomedical sciences, but the primary education objective for the faculty in many PhD degree programs is to prepare students for careers in biology research. We discuss the reasons for this faculty-oriented research focus in Chapter 7. The MD-PhD program integrates a graduate research training program with medical school (3 to 4 years of graduate school plus 4 years of medical school) in a manner that prepares students for careers in research of the cause and treatment of human disease. As discussed, you can take the PhD and MD as separate degrees, but the amount of time involved is increased and the integration of the educational focus of the two degrees may differ considerably as compared with when you take the combined degree program.

■ HOW WILL YOU FUND THE COST OF GRADUATE STUDY?

Frequently, the initial consideration for many students who have been struggling with the costs of a college education is how much will a graduate education program cost? The good news is most PhD and MD-PhD programs in biomedical sciences financially support graduate students throughout their education and research programs. There are two types of funding mechanisms for graduate students: graduate assistantships and graduate fellowships. Graduate assistantships provide a stipend, tuition, and healthcare benefits. They usually have a service requirement, which involves a time commitment to teaching, research, or administration. The amount of time committed to these activities depends on the type of appointment in the graduate school and your graduate advisor. If you are offered an assistantship rather than a fellowship, be certain you understand the time commitment involved in the appointment before you accept the graduate school offer. A graduate fellowship also provides a stipend, tuition, and healthcare benefits but does not have a formal service commitment. In some graduate programs, a fellowship appointment may include teaching responsibilities, but the time commitment is usually less than if you were appointed as a graduate assistant. As we discuss later, teaching responsibilities can be a positive component of your PhD or MD-PhD program, as long as the time commitment does not significantly detract from your research and education activities.

Graduate school stipends

Federal training grants to biomedical graduate school programs and faculty research grants pay for graduate fellowships that include tuition and healthcare benefits and provide a stipend for the student's living expenses. The amounts of the stipends vary depending on the school ($20,000 to $30,000). Government-sponsored fellowships in 2007 from the National Science Foundation provided $30,000 per year (http://blogs.asee.org/fellowships/general-student-fellowships-and-scholarships), whereas stipends funded in 2007 by the National Institutes of Health (NIH) were $20,772 per year (http://grants.nih.gov/grants/guide/notice-files/NOT-OD-07-057.html). Although the stipend will not likely permit an extravagant lifestyle, it is usually sufficient for the basics of life and should keep you from going into debt to complete your graduate school training. Stipends funded by graduate schools usually closely match the federally funded stipend levels. Although federal training grants do not support international students, institutional funds may provide a stipend to international students. You will likely find differences in the stipend levels at graduate schools in different parts of the country. These funding differences may reflect the local cost of living, how competitive the school is trying to be to recruit good students, or the general amount of funding available to the school. It is shortsighted to use the stipend level as a significant factor when you select a graduate school. The education and research components of the program are much, much more important.

MD-PhD student stipends

Similar to PhD programs, fellowships for students in MD-PhD programs usually pay medical school and/or graduate school tuition expenses and health care. Students also may receive a stipend for their living expenses (frequently at levels that fund PhD students at the institution), at least during the PhD component of the program. The most competitive MD-PhD programs have Medical Scientist Training Program (MSTP) grants from the federal government that pay both student stipends and tuition. There is also a program for underrepresented minority students called the Minority Access to Minority Careers (MARC) (http://www.nigms.nih.gov/Minority/MARC). Although this program is mainly for students pursuing the PhD, some MD-PhD students are also included. Alternatively, schools use a combination of internal funds and external faculty research grants to fund their students.

If you believe a MD-PhD program is a way to get a free medical school education, please read further about the structure of the program. Entering a MD-PhD program is not a financial solution for medical students with financial need. This is a very demanding 7-year (or longer) program. This is not a program for individuals interested primarily in becoming a physician and practicing medicine. It is only for individuals who have a very strong dedication to attaining a career that combines both research and medicine. Many MD-PhD programs require incoming students to sign a legally binding document that requires them to pay back medical school tuition expenses and stipends if the student fails to complete the program or completes just the MD degree.

■ WHAT ARE THE EDUCATION COMPONENTS OF A PhD GRADUATE PROGRAM?

Although the structure of individual graduate programs vary, the basic components of the program described here are common to many PhD programs. PhD graduate programs vary in the number of courses required both within the student's area of specialization and in supporting areas. These course requirements (number of credits and specific courses required) are usually completed within the first 2 years of the graduate program. These classes have many similarities to the upper level science classes you took as an undergraduate. It is clearly beneficial to have a strong undergraduate science background to do well in these classes.

Laboratory rotations

In most programs you will spend part of your time during the first year in multiple (two to four) laboratory rotations. During these

1- to 4-month-long laboratory rotations, you work on a research project in a faculty's laboratory. Rotations may start in the summer before you initiate your graduate studies. The primary goal of these laboratory rotations is for you to identify a researcher's laboratory where your thesis research will be done. Thus as you participate in these rotations you evaluate the research opportunities, the scientific process that is used, the educational environment, and the "personality" in several different laboratories. You also learn laboratory techniques and the process of laboratory investigation in an area of biomedical science while doing a limited research project. Faculty also have the opportunity to evaluate your intellectual ability, creativity, personality, and work ethic. Rotating graduate students are frequently involved with an ongoing project of a postdoctoral fellow or senior graduate student in the laboratory. As you take graduate school classes and become exposed to new areas of biomedical research, you may identify new areas of investigation where you might want to focus your thesis research. Laboratory rotations enable you to "try out" multiple potential thesis projects and potentially identify a research project that you could investigate a hypothesis during the next 3 to 4 years of the training program. Once the laboratory rotations are completed, you select a faculty mentor's laboratory where you will do your thesis research during the next 2 to 4 years. Your decision to join a laboratory is only part of the selection process. The selection of a research laboratory also depends on the faculty member's willingness to accept you into the laboratory. This decision will be affected by the faculty member's impression of your academic abilities, level of motivation, and work ethic as well as available space and funding of the laboratory.

Comprehensive examinations

At the end of the second year of graduate school, students usually take comprehensive examinations. The structure of these exams varies at different institutions. The exams may consist of a series of essay exams or a written research proposal that requires you to integrate knowledge presented to you in your graduate school classes and demonstrate your creativity, knowledge of current research literature, and problem-solving skills. Many graduate programs follow up the written exams with oral exams given by a group of faculty who may query you on areas of weaknesses identified in the written exams or in other areas in your area of science specialization. These comprehensive exams can be one of the most stressful events in your graduate program because they combine recall of large amounts of information with creatively integrating the information to solve problems. On a positive side, these exams force you to review the information you have been exposed to in your graduate courses and require you to study areas of science you were unable to get in formal classes. This review fosters long-term memory retention of information that you may need during your thesis research as well as when you are working in your career. To do well in these exams you need to be productive in the classes you take during your first 2 years of graduate school. You will need to learn how to assimilate and creatively apply knowledge as well as enhance your communication skills. You must pass these comprehensive exams to continue in the research phase of the graduate program. Many programs give students two chances to pass the exam. Students who fail the exams are terminated from the PhD program (although they may receive a master's degree).

PhD thesis

After completing these comprehensive exams, you will conduct a literature study and then prepare a written proposal describing a PhD thesis project that will test a question (hypothesis) that you and your mentor have identified. The thesis proposal is reviewed by a group of faculty (thesis committee) who provide advice and critique of your potential thesis research project. During the next 2 to 4 years, most of your time will be spent performing experiments in the laboratory that test the hypothesis of your thesis proposal. During this period you will learn that not all experiments work (frustration) and experience exhilaration from experimentally discovering something that no one else has ever identified! This aurora of new discovery is what motivates and drives scientists. It makes all the work they invested in their research worthwhile.

As a graduate student, you will be involved in additional activities during your thesis research, including participating in laboratory meetings; presenting seminars on the progress of your thesis project to members of the graduate school program or department; working as a teaching assistant in the instruction of undergraduates, graduate students, or medical students; and making presentations in journal clubs that review recently published articles from science journals. You may also be involved in mentoring or directing the research project of an undergraduate student in the laboratory. These activities enable you to develop teaching skills and verbal presentation skills that you will need during your career. Independently acquiring knowledge through reading current published scientific articles allows you to grow and mature as a graduate student without needing the motivation provided by a test or a grade in a course. During this time, you will invest long hours performing laboratory experiments, increasing your knowledge through independent study, and develop your verbal and writing skills. Most graduate students prepare research publications during their graduate years. This is also a time in your life to interact with your graduate student peers, both intellectually and socially. Many of these individuals will continue to interact with you after you complete graduate school and foster your career development. The faculty in your education program also will be long-term assets in helping you be productive during your career. Many faculty consider their students as "family" and will go to great lengths to help them be productive in their careers.

Once your research advisor and thesis committee determine that your thesis research project is complete, you will write your PhD thesis. This thesis document may consist of a series of manuscripts you have published in peer-reviewed science journals or a PhD thesis manuscript that includes descriptions of experiments and conclusions that tested the central hypothesis of the thesis. Once your research mentor and thesis committee members are satisfied that the thesis document is completed, you will defend the thesis during a meeting of your thesis committee. In most cases, a thesis defense is having your faculty members quiz you on the theoretical and experimental aspects of your thesis research. For most students the thesis defense is a very positive experience and an affirmation of the expertise they have acquired in an area of science.

Learning verbal and writing skills

The goals of the PhD training program are to develop the scientific knowledge, laboratory techniques, and creative problem-solving skills to enable you to become a productive research investigator and expand the knowledge base in a defined area of science. Verbal presentation skills also must be developed so that when you become an independent scientist you are able to explain your research observations and conclusions to others. Opportunities may be available to present your research results as a poster or presentation format at regional or national scientific meetings. Although the concept of "selling" your ideas to others may not be consistent with your perception of a scientist, it is a very important factor that contributes to the success of a research investigator. Few of us were born with these verbal skills, and they are developed through practice. You will also develop important personal contacts in the profession, network, and be able to exchange ideas with colleagues in related areas. In addition, you must develop skills in scientific writing to enable you to write competitive grant proposals to fund your research laboratories as well as write manuscripts that will be evaluated for publication in peer-reviewed journals. Publications in peer-reviewed journals are the currency by which research investigators are judged by their peers. Therefore it is important that graduate students both solve problems in the laboratory and have the oral and written communication skills to effectively convey research conclusions and new ideas to other scientists.

■ POSTGRADUATE TRAINING

Do you believe that 4 to 7 years of graduate training is sufficient to prepare someone for a career as a scientist? Well, the answer is probably not. There are many careers, such as teaching positions at some colleges or positions in business or government, which may not require postdoctoral training. However, these positions are also enhanced by postdoctoral training. If your career goal is to become an independent research investigator, graduate school will not be the end of your training program. Once you complete the PhD degree, you can anticipate continuing your training for an additional 3 to 5 (or more) years as a postdoctoral fellow. A productive postdoctoral fellowship increases your future career options. During the fellowship period, you will continue to develop intellectually and increase your skills as an independent investigator in the laboratory of an established investigator. Your postdoctoral research project can extend the research you conducted as a graduate student or can be in a new area of investigation at a different research institution using new research models and techniques. This can be one of the most productive periods in a scientist's career. Postdoctoral fellows work as quasi-independent investigators who are paid a stipend and have the resources of a fully functioning (and funded) laboratory. Salaries for postdoctoral fellows have increased in recent years as a result of a federal mandate. In 2007 the NIH-funded salary for a starting postdoctoral fellow was $36,996 (http://grants.nih.gov/grants/guide/notice-files/NOT-OD-07-057.html). The goal of this training is to publish papers in high-quality science journals, expand your skill and knowledge base, get to know other scientists in your area of expertise by attending national meetings, and develop a research project that you can pursue when you obtain an independent faculty position. At this point you will have invested 4 years as an undergraduate student, 4 to 7 years as a graduate student, and 3 or more years as a postdoctoral fellow. Many students become involved in more than one postdoctoral experience as they change their focus or desire expertise in another area of research.

■ THE LIGHT AT THE END OF THE TUNNEL: SEARCHING FOR A REAL JOB

After this extensive training period, you should be prepared to search for a real job. This is a competitive process where you seek research positions in academic, government, industry, or research institutions. Data from a 2008 article indicated that assistant professors (entry level) earned a mean gross salary of $66,000 per year, whereas full professors earned $115,000 per year (http://www.sciencemag.org/cgi/reprint/304/5678/1830.pdf). Junior research faculty must work hard to obtain grant support to fund their research program. Are research grants easy for junior faculty to obtain? A recent publication (http://www.sciencemag.org/cgi/reprint/298/5591/40.pdf) indicated that in 2001, the NIH gave out 6,636 research grants to investigators but only 251 went to people age 35 and younger. However, some research grants are reserved for new investigators and may provide valuable startup funds. The NIH Academic Research Enhancement Award (AREA) grant (R15) is restricted to individuals who have not been a major recipient of NIH research funds (http://grants.nih.gov/grants/funding/area.htm). Federal funding support for research goes through up-and-down cycles over the years, and it can be a challenge for investigators to compete for funding for their laboratories. However, well-trained investigators who work in a supportive intellectual environment with collaborative faculty get grants and succeed in developing their careers.

It is important for you to not focus on just the negative aspects of this training program and dismiss the career options in science as too difficult to be worth the investment of time and energy. Almost any professional career that you may consider requires extensive post-undergraduate education and training. Similarly, most professional positions require hard work, and there is a high degree of competition for those who want to be at the top of their field. There are many significant benefits to a career in research. Investigators mentor and educate students during their career, creating their legacy by training the scientists, clinicians, and educators of the next generation. Investigators may travel widely throughout the world, write books, and, if entrepreneurial, interact with collaborators in industry to significantly augment their institutional salary. Most importantly, research contributes to developing the world's knowledge base and can be tremendous fun. Finding components of nature that no one else has ever observed provides an indescribable high! Similar to many other professional careers, becoming an independent investigator, college faculty, or selecting another of the many science-related careers demands a significant commitment of time, on-the-job training, and intellectual involvement. If you are interested in a career where you spend only 8 hours per day, 5 days a week at work, then this is not the career for you. However, if you enjoy being challenged intellectually, have a passion for science, enjoy being independent, and have a strong work ethic, then this career path is worth investigating further.

■ HOW DOES AN MD-PhD PROGRAM COMPARE WITH A PhD PROGRAM?

MD-PhD programs combine the information-intensive, people-oriented training of medical school education with the development of laboratory-based, problem-solving education of graduate school. It is possible for you to obtain the two degrees separately, but the combined degree program usually can be completed in a shorter time (7 to 8 years) than if the degree programs are done separately. It is important to understand that MD-PhD programs are organized to emphasize integration of research with clinical medicine. Although the time allocated to the medical school or graduate school programs may be somewhat shortened for students in the combined degree program in comparison with students in separate

PhD programs or MD programs, the MD-PhD academic program requirements are similar to those for students in the single degree programs. In some institutions a small number of students start a PhD or MD program and then successfully compete for a position in the school's MD-PhD program during their first or second year of graduate school or medical school. We previously discussed that education programs that vary from this scientist/physician model. Other programs offer combined degrees in MD-JD, MD-MBA, MD-MPH, and a variety of other master's degrees (http://www.aamc.org/students/applying/programs/start.htm). An unusual variation is the MD-PhD scholars program at the University of Illinois allows a student to complete the PhD in non-science areas such as English or history.

■ STRUCTURE OF AN MD-PhD PROGRAM

During the first 2 years of an MD-PhD program, you usually spend your time taking the medical school curriculum. This is a rigorous information-intensive education program that combines classes in basic science and clinical components. During the first year, most medical schools offer anatomy, histology, and basic sciences classes (such as bioethics, genetics, biochemistry, immunology, neurosciences, and physiology). Some medical schools offer their MD-PhD students specialized graduate-level courses in molecular biology, genetics, biochemistry, and research ethics to students during the first 2 years of the MD-PhD program to shorten the time required to meet the PhD course requirements. Students also may have the opportunity to test out of some basic sciences courses in the medical school curriculum. Many schools encourage their MD-PhD students to do laboratory rotations during the summers before both the first and second years of medical school. This approach enables you to select your research faculty mentor early your training program. The second year of the medical school curriculum frequently emphasizes the pathological manifestations of specific organs or systems (microbiology, hematology/oncology, renal, endocrine systems, respiratory, cardiovascular, nervous systems, gastrointestinal, musculoskeletal, pharmacology).

At the end of the second year of medical school both MD and MD-PhD students complete the U.S. Medical Licensure Examination

(USMLE) Step 1 (http://www.usmle.org/Examinations/step1/step1.html). This examination evaluates whether you can understand and apply important concepts of the sciences basic to the practice of medicine. It focuses the principles and mechanisms underlying health, disease, and modes of therapy.

During the beginning of the third year of the training program, you switch out of the medical school program and into the graduate program where you take graduate courses. In most MD-PhD programs, students take the full graduate school curriculum that includes the program's core science curriculum plus classes in their area of specialization. Some MD-PhD programs also offer courses or seminars specific for MD-PhD students that are designed to increase the integration of science and medicine. The seminar presentations and journal club presentations previously discussed as requirements for the PhD degree are also required for the MD-PhD students during their tenure in the PhD program. After your graduate school class requirements and comprehensive examinations are completed, you usually work on your PhD research thesis for another 2 to 3 years. After completing your PhD degree requirements, you return to medical school for another 2 years to complete your clinical training requirements. During the third year, these clinical training courses frequently consist of a combination of lecture courses and rotations through various hospital and ambulatory services (such as internal medicine, surgery, pediatrics, obstetrics/gynecology, psychiatry, neurology, and family medicine). The fourth year usually consists of taking elective courses selected by the student in specific areas of medicine. Medical school course options and the sequence in which they are offered vary at different medical schools. The AAMC has a website (http://services.aamc.org/currdir/section2/schematicManagerRemote.cfm) that provides curriculum information on courses offered in medical school years 1 through 4 at each U.S. medical school. During the fourth year of medical school, you will take step 2 of the USMLE (http://www.usmle.org/Examinations/step2/step2.html). This exam evaluates if students can apply medical skills and knowledge of clinical science for the care of patients. Finally, after graduating medical school, you will take the USMLE Step 3 examination (http://www.usmle.org/Examinations/step3/step3.html), which is the final assessment of a physician's ability to practice unsupervised medicine.

■ MEDICAL GRADUATE TRAINING PROGRAMS

Similar to the postgraduate training programs for PhD graduates, MD-PhD graduates acquire additional clinical training in a medical residency and fellowship subspecialty programs. The length of these training programs depends on the specialty selected (3 to 8 years). The medical training programs typically require long hours, but the hands-on training experiences are important to educate the physician in the many clinical issues that may be encountered in their clinical practice. Many medical fellowship programs offer MD-PhDs 1 or more years for laboratory-based research that are protected from clinical responsibilities. These research training components can be comparable with a PhD's postdoctoral training program. After completion of these residency/fellowship/research training programs trainees may take American Board examinations to become certified in their area of subspecialization. After completion of the residency/ fellowship training programs, MD-PhDs search for a position in an academic institution, government, or industry that provides opportunities to participate in laboratory research, teaching, clinical research, or translational studies and have clinical responsibilities as well.

■ IS AN MD-PhD TRAINING PROGRAM WORTH THE EFFORT?

If you believe the MD-PhD training program is very long, intellectually demanding, and physically and mentally exhausting, you are correct. There is a very long training period before you get a "real" job that pays more than a stipend-level income. Given that residents and clinical fellows get paid slightly more than subsistence income, it is unlikely you will need to go into debt during your training program. However, training stipends can be challenging when there are a spouse and children to feed, clothe, educate, and entertain. Is this 10- to 17-year education program worth doing? It is worthwhile if you want to be appropriately trained to do competitive clinical research and treat patients. Medical school does not train its students to conduct research like a PhD training program. Medical school education is an information-intensive program to train physicians to diagnose disease and care for patients. Combined degree programs train students to acquire the clinical skills to treat patients as well as the basic science knowledge and problem-solving skills to do laboratory research. PhDs can also participate in research that involves using human diseases, but most graduate training programs do not prepare them to address clinical problems. The combined degree program prepares physician scientists with a unique set of skills, enabling them to have a significant impact on devising new treatments for human diseases or understanding the mechanisms of a disease process. MD-PhDs frequently have a passion for developing new clinical treatments based on their research studies that have increased the understanding of particular areas of medicine. Although this passion is partially based on altruism, for many it comes from their desire for "discovery"; that is, from the personal challenge of solving problems and developing new applications from their work.

MD-PhD salaries

MD-PhD degree training programs can significantly enhance the career options for an individual. Graduates of MD-PhD programs generally are in careers where research and education plays a central role integrated with the care of their patients. Do physician scientists benefit financially in proportion to the effort invested in their education program? If you work in academic medicine, then the answer to the question is probably not. Although physician salaries vary greatly based on their area of specialization, region of the country, experience, and private practice versus academic centers, the average salary in 2006 for internists was $161,000 (Bureau of Labor Statistics). Although MD-PhDs are generally in high demand in academic medicine, they are usually paid less than full-time clinicians. MD-PhDs are trained in academic medical centers, and a recent report indicates 90% of them want careers in academic institutions.[1] MD-PhDs in industry have greater incomes than their counterparts in academic medicine. However, if your career goal is to make money and become wealthy in biomedical science, then you should focus on obtaining an MD degree rather than entering an MD-PhD program and then select a clinical subspecialty that provides significantly more financial remuneration.

■ SUMMARY

1. PhD and MD-PhD programs generally provide financial support for stipend, tuition, and living expenses.

2. PhD and MD-PhD programs are time-intensive and intellectu-ally challenging programs that prepare students for careers as research investigators.

3. The first 2 years of graduate programs encompass didactic classes and laboratory rotations where students "try out" dif-ferent laboratories and research projects.

4. Comprehensive exams, which usually occur at the end of for-mal graduate school classes, are benchmarks where students must demonstrate their intellectual and problem-solving abili-ties in both PhD and MD-PhD programs.

5. Thesis research projects develop a student's scientific knowledge, laboratory techniques, and creative problem-solving skills that enable a student to become a productive research investigator and expand the knowledge base in a defined area of science.

6. MD-PhD programs combine the information-intensive, people-oriented training of medical school education with the development of laboratory-based, problem-solving education of graduate school.

7. MD-PhD programs are organized to emphasize integration of research with clinical medicine.

8. The first 2 years of an MD-PhD program are usually spent tak-ing the medical school curriculum. This is followed by completing a PhD degree (3 to 4 years), and then students re-turn to the medical school curriculum for the final 2 years of their training.

9. PhD programs are followed by additional postgraduate training programs that enhance the student's skills as an independent in-vestigator.

10. Similar to the postgraduate training programs for PhD gradu-ates, MD-PhD graduates acquire additional clinical training in a medical residency and fellowship subspecialty programs.

■ REFERENCES

1. Ahn, J., Watt, C.D., Man, L-X., Greeley, S.A, Shea, J.A. Educating Future Leaders of Medical Research: Analysis of Student Opinions and Goals from the MD-PhD SAGE (Students' Attitudes, Goals, and Education) Survey. Academic Med. 82: 633–645. 2007.

CHAPTER

3

Preparing for Graduate School as an Undergraduate

■ DEVELOPING A PLAN

Now that you have some understanding of the length and rigor of the PhD or MD-PhD training programs and the career opportunities in science, are you still interested in considering either of these as a potential career path? If so, you need to develop a plan for an undergraduate education program that will enable you to meet the graduation requirements of your school and prepare for admission to your advanced degree program. You also need to appropriately sequence the courses to get prerequisites before advanced classes, separate the most difficult courses in your curriculum into different semesters, and take classes that you need to prepare for the Graduate Record Examination (GRE) and Medical College Admissions Test (MCAT) well in advance of the exams. You also should identify the appropriate time for at least one, or preferably multiple, intensive research experience, participate in worthwhile extracurricular activities, and meet the academic entrance requirements of the PhD or MD-PhD programs where you are considering applying. You may consider this suggested plan to be a lot of effort. However, the goal of this plan is to make your application competitive for high-quality PhD or MD-PhD programs. Where you do your advanced degree training may significantly impact your career success, and you should aspire to gain entrance into a very competitive and productive PhD or MD-PhD program. We discuss competitive programs later.

■ UNDERGRADUATE CLASS OPTIONS FOR TRADITIONAL SCIENCE MAJORS

Many students who apply to biomedical graduate degree programs have followed a traditional science major in biology, chemistry, engineering, or other areas of science during the majority of their undergraduate education. During the course of their undergraduate education, these students have time to develop and implement an action plan to enhance their ability to prepare for and become accepted into a competitive graduate or combined degree program. Other students may start their undergraduate education in a non-science field and, sometime before completing their education program, develop an interest in science and decide they want to pursue a career in biomedical science. Once these students become interested in a career in the biomedical sciences, they need to quickly identify the most important things they can do to use their limited time to develop their science background and increase their competitiveness for graduate school. Although both groups of individuals want to accomplish many of the same activities, students in the two groups may develop distinct strategies to give them a competitive advantage in their application to graduate school.

As an undergraduate you should take a science curriculum that fulfills the science requirements of your department, including basic courses in biology, chemistry, mathematics, and physics. In addition, select courses that form a solid academic base appropriate for the graduate area of interest. Classes that include laboratories should be taken whenever possible. Skills such as learning how to use a micropipette, an electronic balance to prepare a chemical solution, or a microscope to count cells are routinely used as tools in a laboratory experiment. These basic laboratory skills and others must be acquired before you become capable of performing complex experiments in a research laboratory. You should select courses that have an investigative component built into the class with ample opportunity to be a creative independent thinker. You should start in your first year of college to consult with a faculty advisor or a professor in your area of interest within the science department to align your educational goals with the available courses in the science departments. This is a great time to take courses in difference areas of science. Although you may find that some subjects are not your favorite, consider that enlightenment to be an additional part of your learning experience. Hopefully you will find many that are interesting and may be considered further when thinking about career options. If you are interested in the MD-PhD program, you should consult regularly with your pre-med advisor about the appropriate pre-med courses and their proper sequence. All medical schools require a year of general chemistry, organic chemistry, general biology, and physics. Most pre-med requirements must be taken before the end of the junior year to allow you to take the MCAT examination at the optimal time. Other required or recommended courses depend on the prerequisites of the medical school of interest. The website of each advanced degree program should be consulted early in your undergraduate studies. Initial choices of potential medical schools can be ascertained by examination of the Association of American Medical Colleges' current *Medical School Admission Requirements*, which can be purchased on the Association of American Medical Colleges' website (www.aamc.org) or may be found in college career counseling centers.

Although most undergraduate schools have a formal advising program, you should seek out faculty with whom you have confidence and can provide good advice about what classes you should take each semester. You need to plan an approach that integrates classes in an appropriate sequence with sufficient rigor and balance. Medical schools and graduate schools want students who can successfully handle the academic demands of their programs and have the stamina and time management skills to be successful. For biology majors, advanced courses in biology should include cell biology, molecular biology, genetics, and, depending on the interest of the student, physiology, neuroscience, immunology, microbiology, or developmental biology. Biology majors should also take general chemistry, organic chemistry, biochemistry, basic physics, mathematics classes through calculus, and statistics. Chemistry majors should include additional courses in basic biology, cell biology, and biochemistry. Engineering majors should take classes in basic biology, cell biology, and biochemistry in addition to meeting their degree requirements in engineering and mathematics. You should have noticed that biochemistry keeps appearing on each of these lists. This is because biochemistry is a very important preparation for both graduate school and medical school curriculums. Some graduate program entrance requirements may include additional advanced biology, chemistry, and mathematics classes. Biomedical graduate programs have a variety of prerequisite science requirements. The website of the graduate program should be consulted for details of the requirements for each school of interest. Science courses should not be taken pass/fail as many graduate or medical schools may interpret a pass as a low C or not count it at all, especially if it is a required course for the graduate program.

■ CLASS OPTIONS FOR STUDENTS WHO SWITCH TO A SCIENCE MAJOR PARTWAY THROUGH THEIR UNDER-GRADUATE TRAINING

When a non-science student switches to pursue a career in biomedical sciences determines if a fifth year or more of undergraduate classes will be needed to complete the graduate school's prerequisite classes. An actual major in an area of science may not be required. Many PhD and MD-PhD education programs have flexible prerequisite requirements and may focus more on the applicant's grade point average (GPA), research experience, GRE or MCAT scores, and/or the comparative ranking of the applicant's undergraduate college. Although graduate programs may advertise that all applicants need to meet their published prerequisites, a phone call to the admissions office may be useful to identify if the graduate program is willing to make exceptions for individual students. If you phone the office, you should identify a compelling set of circumstances that could warrant waiving one or more prerequisite classes before contacting the admissions office. The graduate school may respond that your application will be considered but be less competitive without specific prerequisites. If this occurs, you then have to choose between taking additional undergraduate classes or applying to graduate programs that do not have the requirement. It is important for all students to carefully identify the prerequisites of each graduate program before they submit an application to the graduate program.

■ UNDERGRADUATE GRADES ARE IMPORTANT

You need to excel in your undergraduate studies. Good grades (3.5 to 3.6 GPA at a minimum), especially in science courses, are a prerequisite to being considered for most PhD or MD-PhD programs. PhD applicants should consult the program's website for minimum GPAs. MD-PhD applicants should consult the Association of American Medical College (AAMC) book, *Medical School Admission Requirements* for mean GPAs of accepted MD or MD-PhD students of the schools that they are interested in attending. If for personal reasons you receive poor grades one semester, you have the opportunity to explain the reasons for your temporary slump when you complete your PhD or MD-PhD application. (We discuss this further in Chapter 6.) Similarly, nontraditional students who have a significant difference in grades from early undergraduate years and current grades need to explain the reasons for the differences in their academic productivity. If you have less than a 3.0 GPA, it is likely you will not be competitive for most good PhD or MD-PhD graduate programs. You may want to pursue a master's degree program in an area of biology or biomedical science or pursue a post-baccalaureate program discussed later in this chapter. If your academic record improves, you may become competitive for a PhD or MD-PhD degree program. In addition to working to get good grades in your undergraduate classes, you also should work to develop positive relationships with your science faculty. You need several letters of recommendation from your science faculty for your PhD or MD-PhD application. As discussed later, these letters should come from faculty who know you very well as an individual. You need to demonstrate to the faculty that you are committed to pursuing a career in research and you are a creative, independent learner.

◼ SCHOLARSHIP OPPORTUNITIES FOR UNDERGRADUATE MINORITY STUDENTS

Racial or ethnic minority students who are underrepresented in research or in medical professions are encouraged to consider a career in biomedical research. If you are in this category, many resources are available to you during your undergraduate education. Information concerning these resources as well as important websites can be found at www.aamc.org/students/minorities/start.htm or on the Association of American Medical Colleges (AAMC) Diversity website (www.aamc.org/diversity). Many enrichment programs are available to assist and prepare underrepresented undergraduate students in biomedical careers. Scholarships for undergraduates from underrepresented groups are available to assist students in their undergraduate studies. Two sources that you should examine are the National Institutes of Health (NIH) Undergraduate Program for Individuals from Disadvantaged Backgrounds (https://ugsp.nih.gov/home.asp?m=00) and the Gates Millennium Scholar (www.gmsp.org). Many summer programs are offered for minority students to explore career options (http://services.aamc.org/summerprograms/). Other programs include the Howard Hughes Medical Institute Program (www.hhmi.org/grants/office/undergrad/) and the NIH summer research program for minority students (http://www.training.nih.gov/student/). The website of the Graduate College at the University of Illinois at Urbana-Champaign has an extensive list of fellowships for science/engineering for women and underrepresented minorities (https://www.grad.uiuc.edu/fellowship/category/11).

Students from underrepresented minorities who are applying to medical school can register for the Medical Minority Applicant Registry (Med-MAR) at the time they take the MCAT (http://www.aamc.org/students/minorities/resources/medmar.htm). The Med-MAR circulates information to medical schools and can assist students in finding an appropriate medical school. The post-baccalaureate programs (discussed later) also can help prepare students academically and in laboratory experiences. Pre-med advisors at undergraduate institutions are usually a good source of information on programs and resources for the underrepresented applicant.

■ CAN COLLEGE GRADUATES WITH NON-SCIENCE MAJORS APPLY TO A BIOMEDICAL GRADUATE SCHOOL PROGRAM?

Some students do not develop an interest in a career in biomedical research until they have been out of college for several years. These nontraditional students can pursue a PhD or MD-PhD but may need to take additional courses or, in some cases, retake certain critical courses depending on the length of time since graduation from college. A position as a laboratory technician is an excellent beginning to a career in laboratory research. Post-baccalaureate programs are another training option. If you are one of these nontraditional students, you need to clearly demonstrate your academic ability in science and a strong interest in a career involving research in addition to meeting the course requirements for admission to the graduate program. You may need to retake the MCAT or GRE because most institutions want scores from tests taken in the past 3 years. It is very important for you to obtain individual advice on what classes you may need to take based on your undergraduate record, what academic and work-related activities you have had since you graduated, and what your career goals are. Nontraditional students should contact an appropriate undergraduate science advisor or contact the graduate school or medical school where you plan to apply for advice on how to become a competitive applicant.

■ POST-BACCALAUREATE PROGRAMS

There are a variety of students who are not competitive for PhD or MD-PhD programs because they do not have the pre-med or graduate school academic requirements, lack laboratory research experience, or have a low GPA. This group could also include minority students who do not have sufficient credentials to be accepted into an advanced degree program as well as "nontraditional" college graduates (science or non-science majors) who have worked for several years after obtaining their undergraduate degree before deciding to apply for an advanced science degree program. One option for students who seek to address their academic or laboratory deficits is to participate in a post-baccalaureate program. In general, there are two types of students who may benefit from participating in a post-baccalaureate program: college graduates who seek a career change and science graduates who seek to enhance their application credentials. These 1- to 2-year programs can provide undergraduate or graduate academic education options and may include laboratory research experiences. Some post-baccalaureate programs focus on preparing students for medical school, whereas others prepare students for careers in research or the health sciences. Some of these post-baccalaureate programs are highly selective in their admissions and very productive in helping students matriculate into advanced degree programs, whereas others are less selective in admissions and less productive in their outcomes. A wide variety of post-baccalaureate programs are available, and several websites can provide useful information (http://www.naahp.org/resources_postbac.htm and http://services.aamc.org/postbac/). If you are interested in these programs, examine the details of the program structure and objectives.

Then evaluate how successful the program is in getting their students into advanced degree programs. Consult with your faculty advisor to help you identify the program that will help you meet your educational goals.

■ SUMMARY

1. Where you do your graduate training significantly impacts your career, and you should aspire to gain entrance into a very competitive PhD or MD-PhD program.

2. Start as early as possible during your undergraduate education to develop and implement an action plan to enhance your ability to prepare for a competitive graduate or combined degree program.

3. Many enrichment programs are available to assist and prepare students who are underrepresented in the biomedical sciences.

4. Students should start in their first year to consult with an advisor or a professor in an area of interest within the science department to align their educational goals with the available courses in the science departments.

5. Biomedical PhD and MD-PhD programs have a variety of prerequisite science requirements, and the website of the education program should be consulted for details of the requirements for each school of interest.

6. Good grades (3.5 to 3.6 GPA at a minimum), especially in science courses, are a prerequisite to being considered for most PhD or MD-PhD programs.

7. Students who have been out of college for several years can pursue a PhD or MD-PhD but may need to take additional courses or in some cases retake certain critical courses depending on the length of time since graduation from college.

CHAPTER
4

Undergraduate Research Experience: How to Get Started

One of the most important issues that you must address to prepare for graduate school is gaining research experience. Having at least one extended exposure to inquiry-based laboratory research is essential for applicants to graduate school. Most MD-PhD programs will not even consider an application from a student who has not been involved in substantive laboratory research. It is critical that students demonstrate an interest in research and an obvious commitment to a career that involves research. One way this can be accomplished is by participating in a series of research experiences.

■ RESEARCH EXPERIENCE ISSUES THAT YOU NEED TO ADDRESS

Where do you find a laboratory that will give you research experience? How many experiences should you have and how long do the research experiences need to be? What outcomes do you need from your research experience? Students in large and small undergraduate institutions may have different options for research. If you are a student who is enrolled at a small undergraduate institution, you may have fewer options for research at your school than do students at large universities. In smaller schools, the size of the research laboratories and their research productivity may be less than what can be found at large schools. However, most undergraduate schools have laboratories that can offer you a position to begin to perform research that expands your knowledge of how experimental science is done. Many schools have more opportunities to do field research as compared with laboratory research. You can learn how the process of research is performed by doing either field or laboratory research, and either type of research can be a way to begin your research experience. Research can be performed part time during the school year or during the summer. Some academic institutions such as Oak Ridge National Laboratory (http://www.orau.gov/orise/edu/ornl/default.htm) have semester-long programs where students can conduct a substantial research project and take additional classes.

Finding a good research mentor

How do you determine how many hours per week will enable you to begin to learn how to do research? This question is difficult to answer because there are many so variables. It takes time for you to work with your research mentor to learn the technical skills needed for the project and to develop an understanding of the research process. Importantly, your mentor must involve you intellectually so you understand both the science behind the problem and learn how to critically ask questions that enable you to experimentally address a problem. Your research mentor may be a graduate student, a postdoctoral fellow, or perhaps the head of the laboratory. The quality of the interactions that you have with your research mentor greatly affects your development as a research investigator. What constitutes a good mentor? An article in the science journal *Nature* summarized the personal characteristics of good mentors from a perspective of international students.[1] This article is worth reading for both students searching for their first mentor and for students entering their PhD or MD-PhD program.

Field research versus laboratory research

If your first research experience is in a small laboratory at your undergraduate institution or involves field research, your second laboratory experience should be in a scientifically competitive research laboratory at another larger institution. Your best opportunity to learn the skills of laboratory research is usually found in a federally funded (grants from the National Institutes of Health or National Science Foundation) research laboratory run by an investigator who regularly publishes in peer-reviewed science journals. Faculty publications can be searched for on websites such as the National Library of Medicine's PubMed (http://www.pubmedcentral.nih.gov/) or on similar science databases available at your school's library. You may find useful information about a research investigator by using Google Scholar (http://scholar.google.com). You could consider volunteering in such a laboratory, even if it means initially washing glassware, preparing reagents, or assisting laboratory technicians or postdoctoral students with their research. Your goal should be to get to a position where you can do laboratory research that is focused on experimentally solving a problem and exercising independent creative thinking.

Once you get into a laboratory, even on a volunteer basis, consider the position as a job. Keep a defined attendance schedule, focus on learning the background of the laboratory projects, show enthusiasm and motivation, work hard to learn laboratory techniques, interact in a positive manner with the laboratory personnel, and, most importantly, become intellectually engaged in the research. If your only research experiences are in laboratories that do not have federal funding and do not publish in mainstream science journals, your graduate school application will be much less competitive. In addition, such laboratories may not provide you with an accurate impression of research as a career. There are good reasons for switching to a new research laboratory. You may be able to get a stipend, a better mentor, or perhaps obtain a more independent laboratory position. Having research experience in more than one laboratory can be invaluable in discovering your research interests and developing an accurate perspective of research needed to be a successful graduate student. However, multiple short research experiences could hurt your competitiveness to gain admission to a competitive advanced degree program. When graduate school admissions committees see these brief experiences on your application, they may conclude that you were trying to just enhance your resume.

Financing college versus having time for a research experience

Your class load or employment requirements may constrain your ability to do research during the school year. If so, consider the option of doing research during the summer. Many students have to earn money during their summers to fund their undergraduate education. However, it is very important for you to gain laboratory research experience in a competitive science environment. Therefore you may need to identify an alternative way to finance your education if you spend one or more of your summer vacations in a research laboratory. You may be able to find a laboratory at your school where you can either volunteer or be paid a stipend for working in the laboratory. Many summer research positions are funded. In some instances undergraduate colleges may allow you to continue the summer research during the academic year, which can be very valuable experience. If the options of finding a research laboratory at your undergraduate school are limited, look into Summer Undergraduate Research Experience (SURE) programs, which are available at many research institutions.

■ SUMMER UNDERGRADUATE RESEARCH EXPERIENCE (SURE) PROGRAMS

SURE programs can be found by searching the Association of American Medical Colleges' (AAMC) website (http://www.aamc.org/members/great/summerlinks.htm), the National Science Foundation (http://www.nsf.gov/crssprgm/reu/index.jsp), or websites of individual graduate schools. Many federal programs are available for students (http://www.training.nih.gov/student/index.asp), and state or private research labs such as Woods Hole Oceanographic Institution (http://www.whoi.edu/page.do?pid=7802) also offer summer research opportunities. A current list of many SURE programs is provided in Table 4-1. Most summer programs provide undergraduates a stipend to do 8 to 10 weeks of research in an investigator's laboratory. The stipends are generally not large but should be sufficient to enable you to pay your summer living expenses and, in many cases, have some money for the subsequent academic year. The programs are designed to introduce undergraduates to investigational research. Graduate school–sponsored programs are a recruitment tool for the graduate program and are designed to identify highly qualified students who can be recruited to attend the graduate school. In the SURE program a research investigator helps you identify a problem related to the ongoing research in the laboratory and provide a mentor (technician, graduate student, postdoctoral fellow) to help train you in laboratory skills and the scientific process. They also provide the laboratory resources needed to experimentally address the research problem. SURE programs usually include research seminars, tours of the research facilities, and a chance to interact with a wide range of scientists, graduate students, and other undergraduate students.

Table 4-1 Summer Undergraduate Research Programs

Applicant admission restrictions are identified as well as the location of the research site. Many of these institutions have multiple summer research programs, and each can have different restrictions for citizenship and minority status. This is an incomplete list of SURE programs because they come and go annually as a result of faculty availability and funds to maintain the programs. Most graduate schools and medical schools have some type of SURE program. Search your local school's website for potential positions.

Alabama	**Tuskegee University** Material Science and Engineering Program Restricted to U.S. citizens and permanent residents Tuskegee, AL http://www.tuskegee.edu/Global/category.asp?C=70877 **University of Alabama** Department of Chemistry Tuscaloosa, AL http://www.bama.ua.edu/~chem/undergraduate/summerprogs/ surp.html
Arizona	**Mayo Clinic** Research available at Mayo campuses in Rochester, MN; Scottsdale, AZ; or Jacksonville, FL http://www.mayo.edu/mgs/surf.html **National Institutes of Health (NIH)** The NIH consists of the Hatfield Clinical Research Center and more than 1,200 laboratories located on the main campus in Bethesda, MD as well as in Baltimore and Frederick, MD; Research Triangle Park, NC; Phoenix, AZ; Hamilton, MT; and Detroit, MI http://www.training.nih.gov/student/index.asp **University of Arizona** Summer Research Institute Restricted to students from underrepresented backgrounds with U.S. citizenship, permanent residence, or refugee status Tucson, AZ http://grad.arizona.edu/sri/
Arkansas	**University of Arkansas** Little Rock, AR http://www.uams.edu/pharmtox/surf/

continues

Table 4-1 continued

California	**California Institute of Technology** Pasadena, CA http://www.surf.caltech.edu/
	City of Hope National Medical Center and Beckman Research Institute Duarte, CA http://www.cityofhope.org/education/summer-student-academy/Pages/default.aspx
	Loma Linda University Loma Linda, CA http://www.llu.edu/llu/medicine/cmbgt/undergrad.html
	Occidental College Department of Chemistry Los Angeles, CA http://departments.oxy.edu/chemistry/surp.htm#surp
	Stanford University Restricted to American citizens/permanent residents Stanford, CA http://ssrp.stanford.edu/
	University of California, Berkley Restricted to U.S. citizens/permanent residents Berkley, CA http://mcb.berkeley.edu/amgen/
	University of California, Davis Davis, CA http://www.gradstudies.ucdavis.edu/surp/index.html
	University of California, Irvine Irvine, CA http://www.rgs.uci.edu/GRAD/diversity/surf.htm
	University of California, Los Angeles Multiple programs are available at this institution Los Angeles, CA http://www.gdnet.ucla.edu/asis/srp/srpintro.htm
	University of California, San Diego Department of Pharmacology San Diego, CA http://pharmacology.ucsd.edu/othersurf.php

California	**University of California, San Francisco** Restricted to U.S. citizens or permanent residents San Francisco, CA http://saa.ucsf.edu/summerprograms/ **University of California, Santa Cruz** Restricted to citizens or permanent residents of the United States and its possessions Santa Cruz, CA http://www.chemistry.ucsc.edu/Projects/surf/ **University of Southern California** Santa Cruz, CA http://graddiv.ucsc.edu/prospective/ugradopps.php
Colorado	**University of Colorado Health Sciences Center** Students who have just completed their freshman year or graduated and international students (with student visas) are not eligible for this program Denver, CO http://www.uchsc.edu/gems/
Connecticut	**University of Connecticut Health Center** Farmington, CT http://grad.uchc.edu/internships/intern_intro.html
District of Columbia	**Howard University** Restricted to U.S. citizens or permanent U.S. residents Washington, DC http://www.howard.edu/amgenscholars/
Florida	**Mayo Clinic** Research available at Mayo campuses in Rochester, MN; Scottsdale, AZ; or Jacksonville, FL http://www.mayo.edu/mgs/surf.html
Georgia	**Emory University** Restricted to U.S. citizens and permanent residents Atlanta, GA http://www.cse.emory.edu/sciencenet/undergrad/SURE/SURE.html

continues

Table 4-1 continued

Georgia	**Medical College of Georgia** Restricted to U.S. citizens or international students currently enrolled in U.S. college/university holding a student non-immigrant visa Augusta, GA http://www.mcg.edu/star/session.htm
	University of Georgia Restricted to underrepresented minorities who are U.S. citizens or permanent residents Biomedical and Health Sciences Institute Athens, GA http://www.uga.edu/gradschool/outreach&diversity/surp.html
Illinois	**Argonne National Laboratory** Restricted to students from U.S. undergraduate institutions Argonne, IL http://www.dep.anl.gov/p_undergrad/summer.htm
	Committee on Institutional Cooperation Restricted to citizens or a permanent residents of the United States The CIC is a consortium of 12 research universities Champaign, IL http://www.cic.net/Home/Students/SROP/Introduction.aspx
	Loyola University Health System Microbiology and Immunology Chicago, IL http://www.meddean.luc.edu/lumen/DeptWebs/microbio/Summer/Summer.htm
	Northwestern University Center for Drug Discovery and Chemical Biology Chicago, IL http://www.research.northwestern.edu/cddcb/calendar/undergraduate.html
	Rush University Medical Center Restricted to Rush Medical college students Chicago, IL http://www.rushu.rush.edu/servlet/Satellite?c=RushUnivLevel2Page&cid=1163336425027&pagename=Rush%2FRushUnivLevel2Page%2FLevel_2_College_GME_CME_Page

Southern Illinois University
Carbondale, IL
http://www.siu.edu/~reach/funding.html

University of Chicago
Restricted to students currently studying at undergraduate
    institutions in the United States
Chicago, IL
http://gradprogram.bsd.uchicago.edu/summer_research/
    summer_research.html

University of Illinois
Applicants should be citizens or permanent residents of the
    United States
Urbana-Champaign, IL
http://www.grad.uiuc.edu/EEP/srop/

Indiana

Indiana University
Program in Animal Behavior
Applicants must be a U.S. citizen or permanent resident
Bloomington, IN
http://www.indiana.edu/~animal/REU/

Indiana University
Multiple program offerings
Bloomington, IN
http://www.bio.indiana.edu/undergrad/opportunities/index.html

Purdue University
Multiple programs offered
West Lafayette, IN
http://www.gradschool.purdue.edu/programs/
    summer.cfm

University of Notre Dame
Global Linkages of Biology, the Environment, and Society
Restricted to U.S. citizens and permanent residents
Notre Dame, IN
http://globes.nd.edu/news-and-upcoming-events/reu.shtml

Iowa

Iowa State University
Computational and Systems Biology Summer Institute
Restricted to U.S. citizens or permanent residents
Ames, IA
http://www.bioinformatics.iastate.edu/CSBSI/

continues

Table 4-1 continued

Iowa	**University of Iowa College of Medicine** Iowa City, IA http://www.medicine.uiowa.edu/biosciences/summer/ programs.asp
Kansas	**University of Kansas** Lawrence, KS http://www2.ku.edu/~selfpro/research.shtml
Kentucky	**University of Louisville** Restricted to University of Louisville undergraduate students Louisville, KY http://graduate.louisville.edu/pubs/srop/SROP
Louisiana	**Louisiana State Health Sciences Center** Restricted to U.S. citizens or permanent residents Shreveport Department of Pharmacology, Toxicology and Neuroscience Shreveport, LA http://pharm.lsuhsc-s.edu/pharm/super.htm

Maryland	**Johns Hopkins University School of Medicine**
	Restricted to U.S. citizens and/or permanent residents
	Baltimore, MD
	http://www.bio.jhu.edu/SURE/Default.html
	
	National Institutes of Health (NIH)
	The NIH consists of the Hatfield Clinical Research Center and
	    more than 1,200 laboratories located on the main campus in
	    Bethesda, MD as well as in Baltimore and Frederick, MD;
	    Research Triangle Park, NC; Phoenix, AZ; Hamilton, MT; and
	    Detroit, MI
	http://www.training.nih.gov/student/index.asp
	
	University of Maryland
	Baltimore, MD
	http://medschool.umaryland.edu/osr/summer.asp

Massachusetts	**Boston University** Restricted to U.S. citizens and permanent residents Boston, MA http://www.bu.edu/urop/forstudents/funds/surf/index.html

Massachusetts	**Harvard Medical School** For minority groups that have U.S. citizenship or permanent residency Boston, MA http://www.hms.harvard.edu/dms/diversity/shurpintro.html
	Massachusetts Institute of Technology Restricted to U.S. citizenship or U.S. permanent residency Cambridge, MA http://mit.edu/urop/amgenscholars/
	Tufts University Boston, MA http://www.tufts.edu/sackler/programs/summer.html
	University of Massachusetts Amherst, MA http://www.neagep.org/spur.asp
	Woods Hole Oceanographic Institution Summer Student Fellowships and Minority Fellowships Woods Hole, MA http://www.whoi.edu/page.do?pid=7802

Michigan	**National Institutes of Health (NIH)**
	The NIH consists of the Hatfield Clinical Research Center
	    and more than 1,200 laboratories located on the main
	    campus in Bethesda, MD as well as in Baltimore and
	    Frederick, MD; Research Triangle Park, NC; Phoenix,
	    AZ; Hamilton, MT; and Detroit, MI
	http://www.training.nih.gov/student/index.asp

	University of Michigan Ann Arbor, MI http://www.med.umich.edu/pibs/prospective/rf/summer.htm

	University of Michigan Biological Station
	Biosphere-atmosphere studies in the changing global
	    environment
	Applicants must be U.S. citizens or permanent residents
	Pellston, MI
	http://www.lsa.umich.edu/umbs/umbs_detail/
	    0,2529,11186%255Farticle%255F20657,00.html

continues

Table 4-1 continued

Michigan	**Wayne State University** Smart Sensors and Microsystem Program Applicants must be U.S. citizens or permanent residents Detroit, MI http://www.ssim.eng.wayne.edu/education/ssim_reu_ program.asp
	Wayne State University School of Medicine Applicants must be U.S. citizens Detroit, MI http://gradprograms.med.wayne.edu/sure.php
Minnesota	**Mayo Clinic** Research available at Mayo campuses in Rochester, MN; Scottsdale, AZ; or Jacksonville, FL http://www.mayo.edu/mgs/surf.html
	University of Minnesota Twin Cities, MN http://www.cbs.umn.edu/main/summer_research/

	University of Minnesota
	The Integrative Graduate Education and Research Traineeship
	    Program on Risk Analysis for Introduced Species and
	    Genotypes
	Applicants must be U.S. citizens or permanent residents
	St. Paul, MN
	http://isg-igert.umn.edu/application/default.htm

Mississippi	**University of Mississippi** Oxford, MS http://www.olemiss.edu/agem/SummerResearch/page2.html
Missouri	**University of Missouri** Restricted to U.S. citizens or permanent residents Columbia, MO http://gradschool.missouri.edu/student-development/ emerge/research/
	University of Missouri Multiple programs offered Restricted to citizens or permanent residents of the United States Columbia, MO http://www.lsurop.missouri.edu/summerOther/index.htm

Missouri	**Washington University**
	St. Louis, MO
	http://biomedrap.wustl.edu/dbbs/website.nsf/ced4aeca
	    9da0585a86257000006095ec/206c12732eafe01386257
	    000006167b3?OpenDocument

Montana	**National Institutes of Health (NIH)**
	The NIH consists of the Hatfield Clinical Research Center and
	    more than 1,200 laboratories located on the main campus in
	    Bethesda, MD as well as in Baltimore and Frederick, MD;
	    Research Triangle Park, NC; Phoenix, AZ; Hamilton, MT; and
	    Detroit, MI
	http://www.training.nih.gov/student/index.asp

Nebraska	**University of Nebraska Medical Center**
	International students who are not currently in the United
	    States and on an F1 visa are not eligible for the program
	    because of visa restrictions
	Omaha, NE
	http://www.unmc.edu/dept/summerresearch/index.cfm

New Hampshire	**Dartmouth University** Molecular and Cellular Biology program Restricted to U.S. citizens or permanent residents Hanover, NH http://www.dartmouth.edu/~surf/

New Jersey	**Princeton University**
	Department of Molecular Biology
	Applicants must be U.S. citizens, permanent residents, or
	    foreign undergraduates attending a U.S. educational
	    institution
	Princeton, NJ
	http://www.molbio.princeton.edu/index.php?option=
	    content&task=view&id=321
	
	**Rutgers University/University of Medicine & Dentistry of
	    New Jersey**
	New Brunswick, NJ
	http://rise.rutgers.edu/

continues

Table 4-1 continued

New Mexico	**University of New Mexico/Sandia Laboratories** Center for Micro-Engineered Materials Applicants must be U.S. citizens or permanent residents Albuquerque, NM http://www.unm.edu/~reu/

New York	**Albany Medical College**
	Restricted to U.S. citizens and permanent residents
	Albany, NY
	http://www.amc.edu/Academic/GraduateStudies/Summer
	    Research.html
	Albert Einstein College of Medicine
	Bronx, NY
	http://www.aecom.yu.edu/phd/index.asp?surp
	Brookhaven National Laboratory
	Upton, NY
	http://www.bnl.gov/education/programs/suli.asp
	Cold Spring Harbor Laboratory
	Cold Spring Harbor, NY
	http://www.cshl.edu/URP/
	Columbia University
	Restricted to Columbia and Barnard undergraduate students
	New York, NY
	http://www.columbia.edu/cu/biology/ug/surf/
	Cornell Medical School
	Ithaca, NY
	http://www.med.cornell.edu/education/programs/tra_sum_
	    res.html
	Fordham University
	Armonk, NY
	http://www.fordham.edu/academics/office_of_research/
	    research_centers__in/the_louis_calder_cen/research_
	    opportuniti/
	Mount Sinai School of Medicine
	New York, NY
	http://www.mountsinai.org/Education/School%2520of%
	    2520Medicine/Degrees%2520and%2520Programs/Summer
	    %2520Undergraduate%2520Research%2520Program

New York University School of Medicine
New York, NY
http://www.med.nyu.edu/sackler/programs/
    summer.html

Rockefeller University
New York, NY
http://www.rockefeller.edu/surf/

Sloan-Kettering
Graduate School of Biomedical Sciences
New York, NY
http://www.sloankettering.edu/gerstner/html/
    54513.cfm

University at Buffalo, SUNY
Restricted to U.S. citizens, permanent residents (i.e., holding a
    green card), or those lawfully admitted into the United States
    for the purpose of academic study (e.g., on an F or J visa)
Buffalo, NY
http://www.ribse.buffalo.edu/index.shtml

University of Rochester School of Medicine and Dentistry
Restricted to U.S. citizens, permanent resident aliens, or foreign
    students with visas from their host institutions
Rochester, NY
http://www.urmc.rochester.edu/GEBS/summer-list.htm

Upstate Medical University of New York
Syracuse, NY
http://www.upstate.edu/grad/summer.php

Wadsworth Center
New York Department of Health
Albany, NY
http://www.wadsworth.org/educate/molcel.htm

Weill Cornell/Rockefeller/Sloan-Kettering
Restricted to U.S. citizens and permanent residents who are
    members of underrepresented minority or disadvantaged
    backgrounds
New York, NY
http://www.med.cornell.edu/mdphd/summerprogram/

continues

Table 4-1 continued

North Carolina	**Duke University** Multiple programs offered Durham, NC http://www.biology.duke.edu/undergrad/fellowships.html **National Institutes of Health (NIH)** The NIH consists of the Hatfield Clinical Research Center and more than 1,200 laboratories located on the main campus in Bethesda, MD as well as in Baltimore and Frederick, MD; Research Triangle Park, NC; Phoenix, AZ; Hamilton, MT; and Detroit, MI http://www.training.nih.gov/student/index.asp **National Institute of Environmental Health Science** Applicants must be a U.S. citizen or permanent resident Research Triangle Park, NC http://www.niehs.nih.gov/careers/research/summers/ **University of North Carolina** Department of Pharmacology Chapel Hill, NC http://www.med.unc.edu/pharm/csfp/index.html **Wake Forest University** Winston-Salem, NC http://graduate.wfu.edu/summerprograms/SROPIntroPg.html
Ohio	**Case Western Reserve University** Department of Pharmacology Cleveland, OH http://pharmacology.case.edu/education/surp.aspx **Children's Hospital Research Foundation of Cincinnati** Molecular and Developmental Biology Cincinnati, OH http://www.cincinnatichildrens.org/research/div/dev-biology/train/undergrad.htm **Ohio State University** Restricted to U.S. citizenship or permanent residency Columbus, OH http://www.biosci.ohio-state.edu/~mgmajors/REU/

Ohio	**University of Cincinnati**
	Research Experiences for Undergraduates in Membrane
	    Science, Technology, and Bio-Applications
	Applicants must be U.S. citizens or permanent residents
	Cincinnati, OH
	http://www.med.uc.edu/pharmacology/reu.
	    membrane.science/
	
	University of Cincinnati College of Medicine
	Restricted to U.S. citizens or permanent residents
	Cincinnati, OH
	http://comdo-wcnlb.uc.edu/MedOneStop/Admissions/
	    SummerEnrichment.aspx
	
	University of Toledo
	Toledo, OH
	http://hsc.utoledo.edu/grad/surf.html

Oregon	**Oregon Health Science University** Multiple programs offered Portland, OR http://www.ohsu.edu/xd/outreach/find-highered.cfm **University of Oregon** Eugene, OR http://biology.uoregon.edu/SPUR/

Pennsylvania	**Carnegie Mellon College of Science**
	Pittsburgh, PA
	http://www.cmu.edu/bio/undergraduate/research/
	    SURP/reu.shtml
	
	Drexel University College of Medicine
	Philadelphia, PA
	http://www.drexelmed.edu/GraduateStudies/Summer
	    ResearchOpportunities/tabid/665/Default.aspx
	
	Penn State University, College of Medicine
	Hershey, PA
	http://www.hmc.psu.edu/summerresearch/
	
	Penn State University
	Biogeochemical Research Initiative for Education
	Restricted to U.S. citizens and permanent residents
	University Park, PA
	http://www.ceka.psu.edu/

continues

Table 4-1 continued

Pennsylvania	**Thomas Jefferson University** Philadelphia, PA http://www.jefferson.edu/jcgs/summer_internship_ 2007.cfm
	University of Pennsylvania Philadelphia, PA http://www.med.upenn.edu/bgs/suip.shtml
	University of Pittsburgh Restricted to students currently enrolled as an undergraduate student in an accredited institution of higher education in the United States Pittsburgh, PA http://www.gradbiomed.pitt.edu/summer_surp.aspx
Rhode Island	**University of Rhode Island** Open to undergraduate students currently enrolled in Rhode Island colleges Kingston, RI http://stac.ri.gov/epscor/education/
South Carolina	**Clemson University** Must be citizens or permanent residents of the United States or its possessions Clemson, SC http://chemistry.clemson.edu/undergraduate/SURP/index.htm
	Medical University of South Carolina Restricted to U.S. citizens or permanent residents Charleston, SC http://www.musc.edu/grad/summer/surp/general.html
	University of South Carolina National Science Foundation Summer Research Institute in Experimental Psychology Columbia, SC http://www.cas.sc.edu/psyc/psycugrad/srimain.html
Tennessee	**Meharry Medical College** Neuroscience Program Nashville, TN http://www.meharry-vanderbilt.org/snap.htm

Oak Ridge National Laboratory
Restricted to U.S. citizens and legal permanent residents
Oak Ridge, TN
http://see.orau.org/ProgramDescription.aspx?Program=10055

Vanderbilt University
Programs available in both basic science research and clinical
    research
Only available to U.S. citizens and permanent residents
Nashville, TN
http://www.reliability-studies.vanderbilt.edu/summer_
    research/index.htm

Vanderbilt University
Reliability and Risk Studies program
Restricted to U.S. citizens and permanent residents
Nashville, TN
https://medschool.mc.vanderbilt.edu/summer_academy/

Texas

Baylor College of Medicine
Houston, TX
http://www.bcm.edu/smart/

Rice University
Institute of Biosciences and Bioengineering
Restricted to U.S. citizens and permanent residents
Houston, TX
http://www.nsfreu.rice.edu/

Texas A&M
Multiple programs offered
College Station, TX
http://ugr.tamu.edu/research-experiences-for-
    undergrads-reu/copy2_of_reu

Texas Tech University
Lubbock, TX
http://www.biol.ttu.edu/wiki.aspx?pg=13

University of Texas at Austin
Program in Cellular and Molecular Imaging for Diagnostics and
    Therapeutics
Participants must be U.S. citizens or permanent residents
Austin, TX
http://www.bme.utexas.edu/reu/

continues

Table 4-1 continued

Texas	**University of Texas Medical Branch**
	Restricted to U.S. citizens
	Galveston, TX
	http://www.gsbs.utmb.edu/surp/
	University of Texas, San Antonio
	Multiple programs offered
	San Antonio, TX
	http://www.uthscsa.edu/outreach/summer.asp
	UT Southwestern Medical Center at Dallas
	Restricted to students who are U.S. citizens or possess an
	    F1 visa
	Dallas, TX
	http://www8.utsouthwestern.edu/utsw/home/education/
	    surf/index.html

Utah	**University of Utah** Restricted to Utah residents Salt Lake City, UT http://web.utah.edu/usurp/
Vermont	**University of Vermont** Students must currently be enrolled at any U.S. college or university Burlington, VT http://www.uvm.edu/annb/?Page=summerfellowships.html

Virginia	**University of Virginia School of Medicine**
	Charlottesville, VA
	http://www.healthsystem.virginia.edu/internet/
	    gpo/srip/
	Virginia Tech
	Macromolecular Interfaces with Life Sciences *(MILES)*
	Restricted to U.S. citizens or permanent residents of the United
	    States
	Blacksburg, VA
	http://www.chem.vt.edu/milesigert/Internships.htm

Washington	**Fred Hutchinson Cancer Research Center** Seattle, WA http://www.fhcrc.org/science/education/undergraduates/

	University of Washington Restricted to U.S. citizens/permanent residents Seattle, WA http://www.washington.edu/research/urp/amgen/index.html
West Virginia	**University of West Virginia** Morgantown, WV http://www.honors.wvu.edu/SURE/
Wisconsin	**Medical College of Wisconsin** Milwaukee, WI http://www.mcw.edu/display/docid23576.htm **University of Wisconsin** Department of Bacteriology Restricted to University of Wisconsin, Madison or University of Wisconsin, Milwaukee students Madison, WI http://www.bact.wisc.edu/reu.php
International	**Institut Pasteur** Paris, France http://www.pasteurfoundation.org/zuccaire.html **Weizman Institute of Science** The Karyn Kupcinet International Science School for Overseas Students Rehovot, Israel http://www.weizmann.ac.il/acadaff/kkiss.html

Applying to SURE programs

SURE programs are competitive (require a high GPA) and may be restricted to students who have completed either their sophomore or junior year of college, are U.S. citizens, or are underrepresented minorities. Some of the SURE programs listed in Table 4-1 identify these restrictions. Start early in the academic year to find an appropriate summer research opportunity. Apply to SURE programs in departments where you believe you might have an interest in the area of science, even if you have not yet had a formal course in that area. You should apply to multiple SURE programs in different areas of the country. This is the time in your education to look for a field of science that motivates you, so be flexible when you are selecting programs where you will apply. Discuss potential SURE opportunities with your faculty. They may have personal relationships with faculty who participate in one or more SURE programs or have had previous students participate in a SURE program. Both connections may give you an edge when you compete for a summer research position. Your primary goal should be to obtain research experience even though your research project may not be in an area of your current major interest. Applications to SURE programs have deadlines in January or February. Check the school's website to determine the deadlines. SURE applications usually require a statement identifying your interest in research (see Chapter 6 on writing personal statements), a transcript of your undergraduate grades, and letters of recommendation from one or more of your science faculty (see Chapter 6 on letters of recommendation). Many SURE programs screen out applicants who identify a primary interest in going to medical school. Applicants should be honest and not hide an interest in medical school but also stress their interest in research. So, to be competitive, include in your personal statement the reasons why you want to gain experience from the summer program to enhance your skills and intellectual development for a PhD or MD-PhD program.

Goals of undergraduate research programs

In addition to learning laboratory skills and the process of doing scientific investigation, your undergraduate research experiences should have another important objective—that is, getting a stellar recommendation for your PhD or MD-PhD application from the head of the research laboratory where you are working. You should focus your efforts during each of your laboratory experiences to demonstrate your intellectual skills, creativity, science knowledge, motivation, and work ethic. As we discuss in more detail later in this book, graduate school admissions committee members give priority to students who have positive recommendation letters from established research investigators. You would benefit greatly by contributing enough data to a project so that you become a co-author on a publication in a peer-reviewed journal. You should be aware, however, that most undergraduates do not become authors on a publication. You also may have the opportunity to give a presentation or poster at your college or at a local, state, or national scientific meeting.

Although presentations at such meetings can be intimidating for first-time participants, it is a terrific learning experience and something you can add to your resume.

Summer research programs can be valuable opportunities to introduce you to both the positive and negative aspects of careers in laboratory research. You should take time to reflect on your experience. You may need to look past specific issues related to the research project in which you were engaged or perhaps to one or more of the individuals with whom you may not have had positive interactions. It is important for you to evaluate how you respond to the positive and negative aspects of doing research. You should determine if the scientific process (designing experiments to answer a specific question and possibly discover new insight concerning the scientific issue) that is integral to laboratory research is your passion. It is possible that the experience of a SURE program may show you that research is not what you want to do with your life. If this is the outcome, consider it a valuable learning experience and move on to examine other career options.

■ WHEN DO YOU START PARTICIPATING IN RESEARCH?

Science majors should consider beginning their research experience at the latest by the summer between their sophomore and junior years of college or during their sophomore year. The basic skills of formulating a problem, designing experiments, and critically evaluating experimental results take time to learn. At this point in your education, you have a limited background in science so you need to do a literature search to understand the background issues related to your research project. This results in reading books, review articles, and articles in science journals to understand the knowledge base behind the problem you are trying to address. Read the publications that the laboratory has published and try to understand the research questions that the laboratory is addressing. Talk with members of the laboratory to help clarify issues that you do not understand. It can be beneficial to repeat experiments that others have done to enhance your experimental skills. However, doing experiments that others have done because you are not familiar with the scientific literature can be a great waste of your time and the laboratory's resources. Discussing the published literature with your laboratory mentor and others in the laboratory can enhance your ability to discriminate experimentally sound science from poorly developed ideas. Your laboratory mentor can foster your intellectual development by helping you improve the design of your experiments as well as determine how your experimental results support the hypothesis you are trying to evaluate. The more independent thinking and creativity you can utilize, the more insight you will obtain about a potential career in research.

During the first semester of your sophomore or junior year, look for summer laboratory research opportunities at your school and/or apply to SURE programs at other schools. Several different lists of SURE programs are available in Table 4-1 and in Appendix B. You also should check different graduate school or medical school websites. Most undergraduate colleges have a central location such as a bulletin board where summer research opportunities are posted. Be aware of the application deadline and complete the application in a professional manner. Your competitiveness for a high-quality PhD or MD-PhD program will be enhanced if you can participate in two different research experiences during the summers at the end of your sophomore and junior years of college. Being able to see how research is done in different laboratories can give you a more complete view of research. If you became interested in science late in your undergraduate studies, you also need to have laboratory research experiences. Because your research options are limited by time, it is valuable to find at least one competitive research laboratory where you can spend 3 or 4 months working at least part time on a project or full time in the summer. Because the science process skills you need to develop are the same as those outlined for an undergraduate majoring in science, you need to focus your activities to accommodate the time limitations.

Maintain a journal of your laboratory experiences

Even if you are not required to do so, during each of your research experiences you should write up a detailed account of the laboratory skills you are learning, the background, design, and outcomes of the experiments you are performing and the conclusions of the research project. The journal should be reflective as well as informational. Many undergraduate institutions have undergraduate research symposiums or local scientific meetings where you can present your research in a poster format or give a presentation. You will find that keeping a journal of your laboratory activities greatly helps you prepare for such presentations. Keep a file on the scientific literature you have collected on your project's subject. The journal and the literature reference files also are very useful to you in preparing graduate school (medical school) applications, possible graduate fellowship applications (discussed later), and personal interviews.

■ "PLAN B" FOR STUDENTS WITH NO OR LIMITED RESEARCH EXPERIENCE

If you have been unable to participate in the undergraduate research experiences just described, do you have other options to gain the research experience needed to make your PhD or MD-PhD program application competitive? The best option for you is to work for a year or two after graduation from college as a research laboratory technician to enhance your research experience. You should discuss possible laboratory job options with your faculty as well as go to the human resources office at your institution or other local schools or businesses to seek information about available laboratory technician positions. There are also federal programs such a the National Institutes of Health Post Baccalaureate Intramural Research Training Award Program (http://www.training.nih.gov/student/Pre-IRTA/previewpostbac.asp) or the Centers for Disease Control and Prevention Laboratory Fellowships (http://www.aphl.org/training_and_fellowships/Pages/default.asp) where you could participate in a research laboratory for 1 to 2 years after completion of your undergraduate degree. These programs are designed for students who want to enhance their research experiences before completing an application to medical school or graduate school. Websites describing other post-baccalaureate programs are in Appendix B.

■ EXTRACURRICULAR ACTIVITIES THAT ENHANCE YOUR PhD OR MD-PhD APPLICATION

Although research experiences during your undergraduate years are very important, there are a variety of extracurricular activities that can potentially enhance your application to a PhD or MD-PhD program. Admissions committees for the two advanced degree training programs have different perspectives on extracurricular activities. For PhD programs, research experience is the major activity that enhances your application and you will not be a competitive applicant without significant research experience. Other activities that demonstrate intellectual maturity and curiosity, work ethic, and motivation can have positive effects on being accepted. MD-PhD programs require applicants to have experiences in medically related activities as well as research because you may have to meet both medical school and MD-PhD (graduate school) admissions requirements. For medical schools, participation in multiple healthcare activities is required, and a significant commitment to service activities must be demonstrated. You should directly interact with people who have health problems so you understand the complex personal issues these individuals face as well as appreciate the financial and staffing issues related to the delivery of health care. You also should observe the altruism, empathy, and compassion of healthcare professionals. Activities such as volunteering at a nursing home, hospital, women's shelter, or health clinic enable you to directly participate in, rather than just observe, healthcare delivery. These activities should be long-term rather than short-term experiences. Careers in medicine are not for everyone, and direct experience is the only way for you to determine if this is what you want to do in your career. Medical schools want applicants who are well rounded with a broad academic base and a demonstrated motivation to be a physician. You need to exhibit good time-management skills and a balanced life integrated into a rigorous academic load. Your school's pre-medical advisors should be able to help you identify opportunities for these volunteer experiences in your community.

■ SUMMARY

1. In addition to meeting the graduation requirements of your department, students should take laboratory classes to develop basic laboratory skills, which must be acquired before a student becomes capable of performing complex experiments in a research laboratory.

2. The school website of different PhD and MD-PhD programs should be consulted early in your undergraduate studies to identify required and recommended prerequisite classes.

3. You must excel in your undergraduate courses to have a competitive application for PhD or MD-PhD programs.

4. Having at least one extended exposure to inquiry-based laboratory research is essential for applicants to graduate school and is mandatory for a MD-PhD program.

5. If your first research experience is in a laboratory at your undergraduate institution or involves field research, your second lab experience should be in a federally funded, scientifically competitive research laboratory. This second experience may have to be at another institution.

6. Participation in laboratory-based research programs helps you to understand both the science behind the problem and learn how to critically ask questions that enable you to experimentally address a problem.

7. Participation in summer undergraduate research experiences can significantly enhance your understanding of the research process.

8. You should focus your efforts during each of your laboratory experiences to demonstrate your intellectual skills, creativity, science knowledge, and work ethic to your research mentor.

9. The more independent thinking and creativity you can utilize during your research experiences, the more intellectual development you will acquire and the more insight you will obtain about a potential career in research.

10. During your research experiences you should maintain a detailed summary of your lab skills learned; the background, design, and outcomes of the experiments performed; and the conclusions of the research project. This information will be invaluable in preparing for your PhD or MD-PhD applications.

11. For medical school applicants, participation in multiple healthcare activities is required and a significant commitment to service activities must be demonstrated.

■ REFERENCES

1. Lee, A., Dennis, C., Campbell, P. Having a good mentor early in your career can mean the difference between success and failure in any field. Adrian Lee, Carina Dennis and Philip Campbell look at what makes a good mentor. Nature 447: 791–797. 2007.

CHAPTER
5
Selecting Where to Apply to Graduate School or Medical School

■ **IDENTIFYING AN AREA OF RESEARCH SPECIALIZATION**

During the fall of your junior year of college, you need to focus your efforts on identifying possible schools for your post-baccalaureate education. We will assume that you have taken a breadth of undergraduate classes in biology, chemistry, math, and possibly engineering. These classes, plus your research experiences, should have exposed you to different areas of science and enabled you to decide that research is your primary career goal. You should be able to identify one or more areas of science in which you would like to focus your graduate training. It is important to keep in mind that your area of specialization may change as you take additional classes in graduate school or medical school. Many students select their area of specialization as a result of a positive interaction with a faculty instructor or research mentor, and you will have additional opportunities to be exposed to such key figures in your life during your post-undergraduate training. Most undergraduate institutions have regular research seminars, and these can be very useful in enhancing your understanding of additional areas of science that may interest you. Do not become frustrated because you do not understand the issues discussed at every seminar you attend. Your breadth of understanding different areas of science will improve as your background in science increases.

■ HOW DO YOU IDENTIFY THE GRADUATE SCHOOL/ MEDICAL SCHOOL THAT IS BEST FOR YOU?

You should start to investigate various graduate school options early in your academic career by searching the Internet and talking with your science faculty and the staff of your school's career counseling office. You also should take advantage of opportunities to talk to graduate students or recent PhD or MD-PhD graduates concerning how they selected their PhD or MD-PhD program. Attend presentations or seminars given by graduate school faculty at your institution. You may be able to participate in short informal visits to graduate schools. Such visits can be very useful to introduce you to graduate education options. It is very important for you to understand that there are a number of education programs that can appropriately prepare you for a career in research. At this point in your undergraduate education, your focus should be to identify a small group of PhD or MD-PhD programs that will enable you to meet your education goals.

Because there are a large number of education institutions that have biomedical PhD or MD-PhD degree programs, what process should you use to identify the schools where you will apply? First, you need to identify the criteria that you will use to select a larger group of 10 or 20 schools where you would consider applying. To select these criteria, you should identify your science interests as well as critically evaluate your productivity as an undergraduate student. This initial evaluation is best done in consultation with both your undergraduate science advisor and with your research mentors. As a result of your previous interactions with them, these individuals should understand your strengths, weaknesses, and education goals. You may not receive the same advice from each of these individuals, but you will get additional information that you can use to make an informed, introspective decision. Nearly all academic research programs have strengths and weaknesses in their research. To maximize your career options, you want to set your ambitions on programs that have a group of highly productive research investigators in areas of research in which you have an interest. PhD and MD-PhD programs at these schools also have differences such as their size, the structure of their degree programs, and their record of producing graduates who have successful careers in academia and industry.

■ HOW TO IDENTIFY PhD AND MD-PhD PROGRAMS THAT HAVE STRONG RESEARCH PROGRAMS IN AREAS OF SCIENCE THAT INTEREST YOU

If you think about the advanced science classes that you have had, which of them motivated you to want to learn more about the subject? Which classes were you passionate about? Did you learn about medical or research applications of an area of specialization that interested you? During your research experiences, did the research project that you worked on keep you thinking about new avenues of investigation or possible applications after you left the lab at night? Like many other careers, research productivity is fostered when the area of investigation both excites and highly motivates you. Most graduate school PhD degree programs are in broad areas of science and consist of research investigators who work in narrow areas of specialization. So, the first criteria you initially need to identify when preparing your list of schools where you will apply is what are the areas of science that interest you and which schools have strong programs in these areas? Graduate schools differ in how areas of science investigation are organized. For example, some schools may have an immunology department but do not have a microbiology department, whereas others may have immunology research in the microbiology department. Some schools have a cluster of investigators (program) focused on an area of research even though the faculty are located in different departments. You need to look beyond the university's reputation and evaluate the departments, programs, and individual investigators located at that university. We discuss strategies for making these evaluations later in this chapter.

Useful websites for identifying PhD programs

Use your computer to learn about possible schools that have degree programs in your areas of interest. Start by going to Peterson's website to identify academic institutions that offer PhD degrees in biomedical science (http://www.petersons.com/GradChannel/code/AcdSearchResults.asp?sponsor=1).

Peterson's website enables you to search for institutions that offer PhD degrees in specific areas of science. Books and material in print are less useful for this search because they are usually out of date and do not allow you to search within the material. Most students use the area of the country where the graduate school is located as another criterion for their selection process. After all, you (and possibly your family) have to live at the school's location for at least 5 years. The Peterson's website will enable you to limit your search to institutions in specific areas of the country. The website also includes education program descriptions. Once you narrow your list of possible schools, the Peterson's website provides links to the websites of individual graduate programs. These websites provide additional information on programs of study, financial aid, living and housing costs, location, research faculty, research facilities, life at the academic institution, and e-mail addresses that can be used to request additional information.

Useful websites for identifying MD-PhD programs

A website that provides a list of all MD-PhD programs is found at http://www.aamc.org/research/dbr/mdphd/programs.htm. Although the site does not have specific information about the different programs, links to each MD-PhD website are provided. You should use the current Association of American Medical Colleges' (AAMC) *Medical School Admission Requirements* for advice and information about each medical school (http://www.aamc.org/students/applying/msar.htm). Many MD-PhD programs compete for training grants from the National Institutes of Health (NIH), and institutions that have successfully obtained these Medical Scientist Training Program grants can be found at http://www.nigms.nih.gov/Training/InstPredoc/PredocOverview-MSTP.htm. Medical Scientist Training Program grants are peer reviewed and highly competitive. Institutions that have these grants provide trainees with a wide range of rigorous graduate and medical education training programs. Institutions that have these grants are highly productive research institutions and can provide excellent training opportunities for their students.

Restrictions on funding international students

International students do not qualify for federal or state-funded education MD-PhD programs and may be at a disadvantage when applying to MD-PhD programs unless the institution has private foundation funding. Non–U.S. citizens have few limitations when applying for PhD programs. Although their funding cannot come from federal or state funds, most graduate schools have alternative sources of funding for foreign students.

■ ALIGN YOUR ACADEMIC SKILLS WITH A DEGREE PROGRAM'S REQUIREMENTS

One of the first criteria an admissions committee examines in an applicant's application is their grade point average (GPA) and their Graduate Record Examination (GRE) and Medical College Admissions Test (MCAT) scores. Schools frequently reject your application if you have a minimum GPA or minimum score on these standardized exams. Look at the degree program's average applicant GPA and GRE and MCAT scores (program websites, the AAMC website [http://www.aamc.org/data/facts/start.htm], or phone the admissions office for the information). Do you fit the program's profile? Because there are many options for PhD and MD-PhD programs, take the time to review many programs. For example, at least 118 of 126 medical schools offer MD-PhD programs.[1] If you have a 3.3 GPA and limited research experience, you should strike schools such as Harvard off your list of prospective schools. If you have a low GPA, you might be advised by your faculty advisor to enter a master's degree program before making an application to a quality PhD or MD-PhD program. On the other hand, if you have a high GPA and one or more productive research experiences, you should apply to programs that are doing high-impact research. Be realistic, but do not sell yourself short. You should include degree programs in your list where you fit the entrance criteria and include a few where you may be below the potential cutoff scores. As we discuss later, there are additional criteria used by admissions committees to evaluate prospective students. You may emphasize additional strengths in your application to get the admissions committee to invite you for an interview.

■ HOW TO COMPARE A LIMITED NUMBER OF EDUCATION PROGRAMS

Based on the information we have discussed concerning a degree program's research strengths, you may have shortened your list of potential graduate/medical schools where you will consider applying. Now, how do you compare the programs that remain on your list? PhD and MD-PhD applicants want to use a partially overlapping set of criteria in their comparison, because both want good research options during their graduate school training. Other criteria are unique to PhD and MD-PhD programs, and we consider those later.

Evaluate faculty research productivity as one criterion of an education program's productivity

First, you need to evaluate the research interests of faculty in the department or interdisciplinary program that have research programs in your area of interest. This may be overwhelming if several hundred faculty are listed as department members. Focus your evaluation on the faculty whose research interests align with yours. These individuals will be potential research mentors for you if you are accepted into the program. If there is only one faculty member who does research in areas that potentially interest you, perhaps you should scratch this school from your list. Each school has strengths and weaknesses in its research programs. You need to align your interests with schools that have multiple faculty doing high-impact research in your areas of interest. The admissions committee will likely make a similar evaluation with the information on your application, and it is unlikely they will invite you for an interview if they believe their research strengths do not match your interests.

Look at the program's website to see if there is a list of recent faculty publications. These publications are one measure of the faculty's productivity. Many graduate school programs are not good at maintaining an updated website, so you may need to look up faculty publications in the National Library of Medicine's PubMed (http://www.pubmedcentral.nih.gov/) or in another science journal database in your college's library. Look at the quality of the journals where the faculty publish and the number of their quality publications within the past 5 years. If you are uncertain about the ranking of different science journals, look at Thompson's Journal Citation reports (http://portal.isiknowledge.com/portal.cgi/jcr?Init=Yes&SID=A115J21NPkJKeeOmpMP). Very high-impact journals such as *Science* and *Nature* have impact factors greater than 25, whereas good mid-range journals may have impact factors greater than 5.0. Journals with impact factors of 1 or 2 are much less competitive.

Thus if the faculty have multiple publications in the 5 to 25 range, that suggests they may be productive investigators. If they have publications in journals that rate less than 2, be skeptical about the quality of their research. Although this is not the only measure of journal quality, the frequency that the journal is cited in the science literature is one evaluative approach.

Evaluate faculty collaboration

Another criterion worth evaluating is determining if multiple faculty in the graduate program co-author publications. Researchers are usually most productive, and the people in their laboratories enjoy working in the laboratory, if the laboratory is collaborative with others in the graduate program. You do not want to become part of a laboratory where the head of the laboratory is an isolated "lone ranger." Investigators who work only within their group may also be secretive and even paranoid—a potentially unpleasant environment for people who work in the laboratory. One of the very best aspects of science is collaboration—sharing experimental results and discussing ideas with your peers. Look for evidence that faculty collaborate with each other.

Faculty grant support is another evaluative criterion of productivity

Being productive in science is not only reflected in publications in high-quality journals but also in the ability of an investigator to obtain peer-reviewed grant support. For biomedical scientists, this usually means being funded by extramural grants from the NIH, National Science Foundation, Howard Hughes Medical Institute, or disease-oriented organizations such as the American Cancer Society or American Heart Association. An investigator's grant funding is obviously important to fund laboratory research projects. It also is important to fund graduate student stipends. Currently, NIH funding of research investigator grants is low, and many previously funded investigators are finding it difficult to maintain their extramural funding. This can have a direct effect on graduate students because in most graduate schools, graduate student stipends are funded by the research investigator's grants during the research phase of the degree program. Graduate programs may have alternative sources of income to fund graduate student stipends, but these funds vary greatly in different institutions. Therefore another criterion you should evaluate is the extramural grants (funds from outside of the institution, number of grants, amount of funding) that faculty in the program obtained during the past 5 years. This information is usually not listed in the graduate program's website. Fortunately, the NIH provides websites that enable you to search for the funding history of institutions (http://grants.nih.gov/grants/award/trends/FindOrg.cfm) or for individual investigators (http://crisp.cit.nih.gov/crisp/crisp_query.generate_screen). Faculty tenure is determined, in part, by the number of grants and amount of extramural funding an investigator has obtained.

Tenured faculty in a degree program

Another issue you should consider is how many faculty are tenured (associate professors or professors) or are recent (non-tenured) additions to the department (assistant professors or instructors). Non-tenured faculty have the liability that they may leave (not receive tenure) before the student completes their degree. However, they also may be good student mentors because they have enthusiasm for research and spend a lot of time with individuals in their laboratories. A department or program that has a large fraction of non-tenured faculty may have significant faculty turnover during the period of time that you will be a graduate student.

Now that you have considered faculty interests and productivity, you should evaluate how productive the graduate students have been during the past 5 years. Many programs publish recent student publications on their websites. In good graduate programs, the students are productive and publish in high-impact journals. These programs want to advertise their student's productivity. If the programs do not provide these data, it is difficult to obtain this information by searching PubMed or other science journal databases, and it may indicate a lack of publications by graduate students. If the list is not on the program's website, you can request a list of publications from the admissions office.

How do you evaluate the academic structure of different education programs?

Now that you have addressed the historical productivity of the graduate programs, you should compare the academic organization of the different graduate programs. There are philosophical differences between research faculty as to the number and breath of classes graduate students should take as compared with how quickly they should begin their lab research. The graduate system in England is one extreme example because students take very few formal classes and start laboratory research very early in their training program. This extreme does not exist in the United States. Your graduate training program may be the last opportunity to obtain formal education in fields directly related to your area of specialization. You need breadth in your education to be able to quickly change research directions in your career. Graduate school classes also introduce you to new areas of specialization. You may change your area of specialization as a result of being exposed to a new and exciting area of science. Alternatively, you may encounter a faculty instructor in one of your classes who relates exceptionally well to students. As a result of this interaction, you may decide this is the person with whom you want to do your PhD thesis. Take advantage of the educational opportunities available in your graduate program and learn the depth and breadth of a field of science.

As discussed earlier, many PhD students complete their class requirements during the first 1 or 2 years of graduate school. MD-PhD and PhD students usually have very similar course requirements. Does the program have a core curriculum that all students must take? A core curriculum is a set of courses in a variety of disciplines that the school believes all students should experience. In recent years, many graduate schools have adopted a core curriculum so that their students get a good foundation in biomedical science. When you are comparing different advanced degree programs, look at the list of classes available to students—both required and optional classes. If you decide to pick an area of specialization where you may benefit from classes in another graduate program at the institution or at a neighboring institution, how difficult will it be for you to obtain these classes? The program should be designed to help you meet your educational goals, rather than a one size fits all structure. Talk with your undergraduate faculty mentors about their opinion of optimal graduate school curriculum programs before you decide what may be best curriculum organization to enable you to meet your educational goals.

Education program issues specific to MD-PhD applicants

The evaluation issues just discussed for PhD applicants are also applicable for MD-PhD graduates. In addition, the AAMC has a website (http://www.aamc.org/data/gq/start.htm) that can be used by MD-PhD applicants to obtain information about specific medical school programs. The site has responses to a survey of graduating medical students concerning the student's medical school experiences, student support programs, and potential problems, including mistreatment. Where have the graduates gone for additional clinical training? This information should be available from the admissions office of the program. It may be difficult for you to comparatively evaluate the quality of different clinical training programs, but your pre-med advisor should be able to help you determine how research intensive the medical institutions may be. Another selection determinant that is specific to MD-PhD rather than PhD programs is that some state-funded schools select a large fraction of their students from resident (in-state) applicants. State residency is usually very restricted for students pursuing a medical degree at state-funded medical schools but may be waived for students in the MD-PhD program. Consult the *Medical School Admission Requirements* book described and look at the state residency of the students accepted to each program. Private medical schools usually do not have state residency requirements. Check the websites of specific MD-PhD programs (or medical schools) to determine if this bias might enhance (or reduce) your chances of being selected. The AAMC has another website (http://www.aamc.org/data/facts/start.htm) with useful information concerning the GPA scores, MCAT scores, ethnicity, sex, and state of residency of recent medical school applicants. State residency requirements are not a factor in admission to most PhD programs.

MD-PhD programs differ in how basic science classes are integrated with medical school classes during the first 2 years of the program. Classes for MD-PhD students may be on a fast track as compared with classes for graduate students. Are most of the initial science courses for MD-PhD students taken with medical students or are the courses limited to graduate students? General medical school science classes tend to be less rigorous than graduate school science classes or classes provided only to MD-PhD students. Is there a component of the MD-PhD curriculum that is dedicated only to MD-PhD students? These could be basic science classes during the first year of the program or classes that help the student transition from the research program back to the medical school curriculum (usually in the fifth year of the program). It also is important to understand how flexible the program is concerning class requirements. MD-PhD students should discuss the quality of the different medical school clinical education programs with their pre-med advisors. The quality of your clinical training should be a major consideration when you select where you plan to apply for your MD-PhD training. Your pre-med advisor also should be able to give you information concerning how effectively the research component of the program is integrated into the medical school's education program.

Curriculum in a medical school can change when a new administration (medical school dean) is hired to direct the education program. The new education initiatives initiated by a new dean can be progressive or, in a few cases, follow fads rather than academic merit. You need to proactively evaluate the structure of the medical school curriculum in the schools where you plan to apply. This is best done in collaboration with your faculty advisor and research advisor. AAMC has a website (http://services.aamc.org/currdir/section2/schematicManagerRemote.cfm) that provides curriculum information on courses offered in medical school years 1 through 4 at each U.S. medical school. You should examine the curriculum in the schools that interest you. Schools that have a nontraditional curriculum should be evaluated carefully. Similarly, schools that minimize their basic science classes for medical students may not provide MD-PhD students with the background they need for competitive careers in research. Do not assume that the "newest" curriculum adopted by a particular medical school is based on data from peer-reviewed education analyses. Medical schools hire new deans for a variety of reasons, and one reason can be related to implementing the latest educational fad into the medical school curriculum. Medical schools have to undergo a reality check from the Liaison Committee on Medical Education (http://www.lcme.org/standard.htm) accreditation reviews, but these generally occur only every 8 years. These accreditation reviews can result in the medical school dean being removed if the school fails to provide a curriculum structure that adequately supports the preparation of the medical careers for the school's medical students. You cannot assume that a "new curriculum" engaged by a particular medical school will provide you with the educational experience that you need to advance your career. You should critically evaluate each medical school where you plan to apply for an educational program that is suited for your individual needs.

■ SUMMARIZING THE INFORMATION YOU HAVE COMPILED

Now that you have assembled information on different aspects of multiple education programs, how do you comparatively rank the schools? Box 5-1 provides a checklist for helping you select to which schools to apply. After you have reviewed the collected information, this is the time to solicit additional advice from your faculty, research mentors, and recent graduates of PhD or MD-PhD programs concerning their perception of the quality, research productivity, and educational effectiveness of different graduate schools (medical schools). You need to define your educational goals with your advisors. You also should review with them the list of graduate programs you are considering and your rationale for selecting each program. Do not be surprised if you get a variety of responses from your advisors—there are bound to be differences of opinion about education programs. Some advisors may be well informed about different programs, whereas others will have a more narrow perspective. Take the information you get from your advisors and the information you have collected from the questions just discussed and use it to make a decision about where you are going to apply (see Box 5-1). Consider applying to a minimum of five different programs. Be realistic; do not apply to a large number and then not have time to adequately fill out the materials or devote the attention that each application requires. Your final list should include programs that you believe are consistent with your academic abilities as well as programs that may exceed your abilities. You need to compile all of this information well in advance of graduate school/MD-PhD program admission deadlines. You still have to write your admission essays, get transcripts sent to each of the programs, and ask faculty at least a month in advance to write letters of recommendation for you.

■ SUMMARY

1. Select a group of 10 or 20 PhD or MD-PhD programs where you will consider applying based on your science interests, your productivity as an undergraduate student, and advice you obtain from both your undergraduate science advisor and your research mentors.

Box 5-1 Checklist for Determining Where to Apply

Schools you are evaluating should get 1 (low) to 5 (high) points for each issue listed below. Include input from your academic and/or research advisors when you evaluate schools. The more information, the better!

Points

Number of research faculty (in your area
of specialization) _____

Quality of research faculty (in your area of
specialization) _____

   Quality of publications _____

   Publications co-authored within department/
program _____

   Investigator's extramural research grants _____

Graduate student productivity (publications) _____

Curriculum options and flexibility _____

Traditional curriculum vs. "new" curriculum _____

Core curriculum for all students _____

Location _____

Cost of living _____

Source of stipend _____

Department/program's academic reputation _____

Student diversity _____

Program's average GRE/MCAT score _____

Average number of years to complete degree _____

Advisor's assessment of educational program _____

For MD-PhDs _____

   In-state advantage/disadvantage _____

   Medical Scientist Training Program grant in
program? _____

   Quality of medical school education _____

Total points: _____

2. Use the Internet to identify PhD or MD-PhD programs that have productive (high-impact publications) and extramurally funded research investigators in the area of specialization(s) you believe you have an interest.

3. Comparatively evaluate graduate/MD-PhD programs by determining if graduate students co-author publications in quality science journals and continue their education in postdoctoral fellowships or research intensive clinical fellowships.

4. Evaluate the availability of graduate classes and flexibility of the education curriculum in each school. The education program should be designed to help you meet your educational goals, rather than a one size fits all structure.

5. MD-PhD students should discuss the quality of the different medical school education programs with their pre-med advisors. The quality of your clinical training should be a major consideration when you select where you plan to apply for your MD-PhD training.

6. Review your compiled data with your faculty advisor and research mentors and use their advice in making your final selection of at least five schools where you will apply.

7. You cannot assume that a "new curriculum" engaged by a particular medical school will provide you with the educational experience that you need to advance your career. You should critically evaluate each medical school for an educational program that is suited for your individual needs.

■ REFERENCES

1. Ahn, J., Watt, C.D., Man, L-X., Greeley, S.A, Shea, J.A. Educating future leaders of medical research: Analysis of student opinions and goals from the MD-PhD SAGE (Students' Attitudes, Goals, and Education) Survey. Academic Med. 82: 633–645. 2007.

CHAPTER

6

The Application Process for PhD and MD-PhD Programs

■ OVERVIEW

There are significant differences in the applications for PhD and MD-PhD programs. PhD applications are sent directly to specific graduate schools, and each school may have differences in deadlines for receipt of applications, academic prerequisites, and application components. In contrast, MD-PhD programs have one primary application that is submitted to the American Medical College Application Service (AMCAS) (http://www.aamc.org/students/amcas/start.htm). This centralized service provides the primary application that is used for all MD-PhD programs. Once AMCAS has a completed application, they forward it to the schools you have identified in the application. The individual schools evaluate your application and, depending on their admission criteria, decide if your application will be considered further or rejected. If you pass this selection evaluation, the school will request that you submit a secondary application and have recommendation letters sent to the admissions office of the school.

■ APPLICATIONS TO PhD PROGRAMS

Most applications for PhD programs consist of six components:

1. An application fee.
2. An official transcript of courses from each academic institution you attended.
3. Graduate Record Examination (GRE) scores.
4. A personal statement that includes a description of your involvement in research and your education goals.
5. Multiple letters of recommendation.
6. Applicants for whom English is not their native language and who have not attended an English-speaking undergraduate institution must provide the results of the Test of English as a Foreign Language (TOEFL).

Transcripts

Most PhD programs do not have a transcript form but rather accept official transcripts from undergraduate school registrars. In your application you will list all the institutions you have attended and provide the attendance dates. You must provide official transcripts from each institution you attended.

Graduate Record Examination

The GRE General Test currently consists of two 40-minute sections on verbal reasoning, two 40-minute sections on quantitative reasoning, and two 30-minute sections on critical thinking and analytical writing. The examination undergoes periodic revisions, so you should check the GRE website (www.ets.org/gre) for the current structure of the examination. This is a computer-based exam given on 30 different test dates at 3,500 test centers worldwide. Before the exam dates, discuss with your faculty advisor when is the optimal time for you to take the exam based on the classes you have taken. Usually, previous performance on the ACT or SAT can be used as a predictive factor for applicants as they prepare for the GRE and can be used to determine how much preparation you may be require. Sample questions, test preparation strategies, and more detailed information about the examination are available on the GRE website. The GRE tests basic skills and is not specific to an area of science specialization. Some graduate programs either require or encourage you to also take one of the GRE subject tests that are given in eight specific fields of study:

1. Biochemistry, cell and molecular biology
2. Biology
3. Chemistry
4. Computer science
5. Literature in English
6. Mathematics
7. Physics
8. Psychology

You should take a subject test in the area that you have majored in, not your undergraduate minor, to maximize your exam score. For example, biology majors should take the biology subject exam rather than the biochemistry, cell and molecular biology exam. Check the website of the graduate schools where you plan to apply for potential GRE subject exam requirements. The subject exams are given three times each year. You should not attempt to take the subject test and the general test on the same day. To adequately prepare for these exams (especially the subject tests), you should begin several months before the exam date. Take practice tests on the computer to acquaint yourself with the format. The GRE adjusts the difficulty of the questions during the test if you are getting questions correct. Talk with your science advisor to determine if you have had the science classes that cover the material in the subject exam. If not, you will need additional time to prepare for the test. There are books (see Appendix B) available to help you review relevant materials in preparation for the exams.

If you do not do well on one or more of the GRE exam sections, you can take both the general and subject tests more than once. Because these are computer-based examinations, you will find out your scores on all but the analytical writing component shortly after taking the exam. Students whose primary language is not English typically have lower scores in verbal reasoning but may do well in the other exam sections. For such students, graduate schools may require a minimum score on the TOEFL examination and, if the verbal scores are low on the GRE, may reexamine the TOEFL scores.

Although one recent research study suggests a correlation between GRE scores and academic productivity in graduate school, graduate school admissions committee members have widely different opinions concerning how well GRE scores predict research productivity in graduate school.[1] Thus the composition of the admissions committee can affect how much the GRE scores may impact your admission success. Some programs may place limited emphasis on your scores unless the scores are very low (less than 30th percentile). Some programs do not accept applications that have GRE scores below a defined minimum score. The minimum scores may be posted on their website or can be obtained if you phone the admissions office. Some undergraduate education programs rely heavily on essay exams rather than the multiple choice exam format of the GRE exams and such students may underperform on GRE exams. In addition, if you received your baccalaureate degree several years before taking the GRE exam, your GRE scores may not be predictive of future performance in graduate school. Thus if your scores on one or more GRE exam sections are poor while your grade point average (GPA) is high, you should provide an explanation in the personal statement of your application as to why you believe the GRE scores are not a valid representation of your academic abilities.

Letters of recommendation

These letters can significantly affect your ranking by graduate school admissions committees—positively or negatively. We previously discussed in Chapter 4 how you should foster interactions with your science faculty and research mentors so they are aware of your intellectual abilities, creativity, lab skills, leadership roles, and motivation for a career in research. The letters they write for you must indicate how they know you and clearly indicate that you have a strong commitment to graduate study. You must have letters of recommendation from all your research mentors. Failure to include letters from one or more research mentors may make your application noncompetitive. If you do not have letters from all your mentors, you should provide an explanation for the absence of the letters in the personal statement of your application. Letters from your research mentors should be signed (or cosigned) by a faculty member rather than a postdoctoral fellow or graduate student (even though they may have been your direct lab mentor). Generally, it is recommended that you sign the waiver giving up your right to read the recommendation. The admission committee usually gives more weight to letters where the student has signed the waiver. If you request a letter from one of your instructors, be certain you got an "A" grade in their class and preferably select classes that had a lab component. It is counterintuitive to have one of your faculty describe you as a terrific student in a course that you received a "C" grade. Do not request letters of recommendation from non-science faculty, friends, or family

Faculty who write letters of recommendation have higher credibility with admissions committee members if they are productive research investigators, have previously written letters for students who have done well in the degree program, or who have extensive experience mentoring undergraduate students that have become productive PhD or MD-PhD students. Do not request letters from faculty who have not had you in class unless they were your advisors and know you well. Do not just select faculty who are prominent unless they really know you. Ask the faculty members if they are willing to write a letter of recommendation for you before giving them the form. Give the faculty member plenty of time, 3 to 4 weeks at a minimum. It is usually very helpful for you to meet with the faculty member and discuss your interest in graduate school, your interest in research, your goals, and the contents of the personal statement in your application (Box 6-1). Faculty will be able to write a more comprehensive letter if you provide them with a copy of your application with your personal statement.

You should review the issues listed in Boxes 6-1 to 6-3 and use the information to critically evaluate your interactions with faculty advisors, research mentors, and laboratory coworkers. These are the issues by which you will be assessed by admission committee members, so use some introspection to evaluate how you perform and interact with others.

Personal statement

A personal statement, which is part of the application, is your opportunity to set yourself apart from other applicants and identify your major talents, experiences, and interests. This is a very important component of the application, and you should invest significant effort to write an effective statement. It should describe your intellectual development, your interest in research, your education goals, what motivates you to become a scientist, and why you are applying to the school. It also should describe how your research interests align with the research strengths of the graduate

Box 6-1 Notes to Faculty:
Letters of Recommendation

The comments faculty write in their letters of recommendation can significantly affect the outcome of a student's application to an advanced degree program. Admission committee members closely read recommendation letters (and, for better or worse, frequently try to read between the lines). Faculty should consider the following issues when writing letters of recommendation:

- If you, as an instructor, cannot write a positive recommendation letter, decline the student's request to write a letter. A faculty member does not want to get a reputation for making negative comments about their students and does want to act in the best interest of their students. If you decline to write a letter, it would be helpful to work with the student to find another faculty member who will provide the recommendation.
- Describe how well you know the individual student. How long and in what context?
- Rank the student in comparison with previous students in lab or class. Compare the student with students who have successfully gone on to careers in research.
- Describe the student's scientific background and academic potential.
- Acknowledge if the student has potential negative characteristics such as a quiet demeanor, English language deficiencies, or poor communication skills. Directly address if these issues affect their productivity, academic ability, or interpersonal relationships. Cite examples concerning how the student has learned how to overcome or compensate for the characteristics. If not directly addressed in reference letter, these issues may negatively affect outcome of the student's interview.
- Identify the student's creativity, work ethic, maturity, and level of motivation.
- How has the student demonstrated a commitment to pursuing a career in research?
- Does the student have sufficient research experience to understand what a career in research involves?

- Is the student an independent learner who is focused on becoming educated rather than just a student who works for high grades?
- Does the student work at his or her full potential? Does the student have the organizational skills, perseverance, and stamina to succeed in an advanced degree program and in a research career?
- There is inflation in the descriptive terminology used by mentors to characterize students. Comments including good or very good suggest modest support of the applicant and may be viewed as negative.
- Does the individual have good critical thinking skills with the ability to apply and interpret concepts presented in class or in the laboratory?
- Does the individual work well (cooperatively) with others in class or the laboratory?
- How has the student's extracurricular activities contributed to the development of the student's leadership skills or understanding of his or her career options?
- Identify major personal obstacles that the student has encountered in his or her life and describe how he or she overcame the obstacle.
- Do not ask your laboratory staff to write letters of recommendation for students in your laboratory. These letters may be dismissed by admissions committees. You should take input from individuals in your lab who may have directly interacted with the student and then write and sign the recommendation letter.

FOR MD-PhD STUDENTS

- How does the student handle a science intensive curriculum, show evidence of compassion, and exhibit a strong motivation for a career involving research and medicine?
- Omissions can be as important as what is included in the recommendation. Applicants may be negatively evaluated by admissions committee members if key issues are omitted from the recommendation letter. Some program directors may phone faculty for clarification of incomplete recommendation letters. Have a smile on your face when you describe the student's abilities.

Box 6-2 Notes to Faculty:
Research Mentors' Letters of Recommendation

Your recommendation letters will be examined for comments cited in the sections on letters of recommendation given previously and the additional components listed below:

- Outline the mentor's research experience (length and depth) and research productivity.
- Describe the mentor's versus student's role in defining the student's lab project objectives.
- Identify the student's ownership of and individual contributions to the research project (independence).
- Does the student have good laboratory skills ("good hands," experimental design, techniques)?
- Identify the student's understanding of technical and theoretical aspects of the project (technician vs. scientist).
- Provide examples of how the student appropriately dealt with a project's obstacles and experimental setbacks (tenacity).
- Was the student a technical resource for others in the lab?
- Describe how the student interacted with others in the research laboratory (collegiality, cooperation, contributions to team efforts).
- Does the student exhibit good communication skills (written, oral)? Can they discuss their research in a coherent, organized manner?

Box 6-3 Notes to Faculty:
Advising Students to Secure Quality Letters of Recommendation

There are a variety of ways to help a student obtain quality recommendation letters. One approach is to help the student provide an organized summary of their credentials to the faculty who will write letters of recommendation for the student. The first step in this process is to interact with the student to identify a series of "talking points" for the student's application. This list may include a summary of their academic and research strengths, addressing many of the issues presented in Boxes 6-1 and 6-2. It also should describe a summary of their research experiences as well as their education, research, and career goals. It is helpful if the student has prepared a draft of their personal statement describing their background, intellectual development, and research experiences. It can be beneficial to have them identify negative issues that they have encountered and overcome. Have them discuss how they overcame the setback and grew from the experience. It also may be helpful if you have independently prepared a list of "talking points" that you believe accurately identify the student's strengths. When you meet with the student, the bidirectional discussion can lead to a valuable clarification of the student's strengths and weaknesses. After discussing these issues with the student, you can make suggestions for revisions of the student's "talking points" that focus the student's academic and personal strengths while presenting potential negative issues in a positive manner.

Discuss with the student how to select potential faculty who will write the letter of recommendation. Stress the importance of the letter of recommendation and the areas the faculty member will be asked to address (Box 6-1). Listen to the student and ask questions to allow the student to select the best faculty who will cast them in a positive manner. Reinforce with the student the need to give faculty adequate time and information to allow the faculty member to write a quality letter of recommendation.

It is then useful if the student subsequently meets with each of the faculty whom the student has asked to write letters of recommendation. The student can present them with their written "talking points" and a written summary of their research, education, and career goals. The faculty will benefit from having an organized written description of the student's research experiences, strengths, and goals when they write recommendation letters. Students should be encouraged to provide each letter writer with an addressed stamped envelope and clearly indicate the deadline. Students should send a thank you note to the faculty member after the letter has been written and inform the faculty about the outcome of the application.

program. You should demonstrate to the admissions committee that your research experience has given you an understanding of the positive and negative aspects of research and that you understand what a research career entails. The document should demonstrate your intellectual curiosity, your persistence when confronted with adversity, and your motivation to be a successful research investigator. Describe how your undergraduate education has stimulated your scientific interests and given you the background to understand the issues that you want to address in your graduate research program.

Your initial paragraph is important to get the attention of the admissions committee members. You should use it to identify who you are, to introduce the main ideas of your personal statement, and to provide direction of where the other components of your statement will go. It is easiest to write this after you complete the other components of your statement. The body of the statement should tell a story incorporating the components we discussed previously. Be specific about your experiences and what they taught you. Avoid writing broad generalizations. Write in a manner that demonstrates confidence without arrogance. Take enough time to make this a polished document.

You should describe in chronological order each of the research experiences in which you have participated and document how your interest in research developed as a consequence of your experiences. For each of your research experiences, identify the role you played in the project (e.g., assistant to a graduate student or postdoctoral fellow, independent project leader supervised by the head of the lab), a brief description of the project's background, the research project's objectives, your intellectual contributions to the project (including obstacles you encountered and overcame), and a description of the project's conclusions. The notes you made during each of your research experiences (discussed in Chapter 4) can greatly help you prepare this section of the application. You should also include a statement concerning how each research project or internship contributed to your education, career path selection, and development as a scientist. Provide a list of presentations you may have made on the project and include copies of any abstracts or publications that came from your projects. You also could identify how your scientific interests identified during your research experiences align with the strengths or research interests of the graduate faculty in the graduate program.

You should include a description of significant work or extracurricular activities that may have enhanced your intellectual development and leadership skills as well as science or non-science awards or honors you have received. Unlike medical school admissions committees, graduate school admissions committees are primarily interested in your science-related activities. Note any postgraduate fellowships such as a National Science Foundation (NSF), Fulbright, or Rhodes Fellowship that you are currently pursuing. You must convince admissions committee members that you have a passion for research, a strong commitment to graduate training, experienced the process of research (testing an experimental hypothesis), understand from your laboratory experiences what is involved in a career as a research scientist, the intellectual ability to excel in graduate school, and experienced both the positive and negative aspects of laboratory research.

Addressing potential shortcomings in your application

If you have encountered difficulties in your personal life, it may be of value for you to briefly describe these difficulties in your personal statement, how you overcame the difficulties, and how this process altered your life. If you have had poor grades during a defined period of your undergraduate education, you should explain why this occurred and how you overcame what caused the poor grades. Similarly, if you have poor scores on a subset of the GRE or Medical College Admission Test (MCAT), you should explain why you believe these scores do not represent your academic abilities. Always attempt to turn a negative into a positive. Sometimes a well-written explanation can address how you grew from the experience and are a better applicant because you have dealt with that experience. However, if you fail to address potentially negative issues in your personal statement, grades, or standardized exams, it is likely your application will be judged negatively.

Once you have a final draft, read the statement carefully to make sure it is devoid of spelling errors and grammatical mistakes. It is usually valuable to have your faculty and research advisors read a final draft of your personal statement and make suggestions for modifications. This statement is a very important component of your application and it is important that it reflects who you are and what you have accomplished.

TOEFL examinations

The TOFEL test is a prerequisite for admission to most graduate programs in the United States for individuals for whom English is not their native language. It also is used by government, licensing, certification agencies exchange, and scholarship programs. The TOEFL examination is designed to evaluate how well you read, hear, speak, and write English. The examination is given as either an Internet-based test or a paper-based test. The Internet test is the primary examination format. However, because the Internet-based examination is not available in all areas of the world, the paper-based examination is available as a replacement. The test is offered 30 to 40 times per year and is given in 180 countries at locations that can are identified on the TOEFL website (http://www.ets.org/bin/getprogram.cgi?test=toefl&redirect=format). The examination is divided into four sections and can be completed in less than 5 hours. The TOEFL website provides practice questions and preparation advice. Commercial books and multimedia programs are also available to help you prepare for the examination (see Appendix B). You can sign up to take the examination on the TOEFL website. TOEFL provides your examination score to you and the institutions you identify 15 business days after you take the examination. Each academic institution sets its own minimum score that it will accept from its applicants. You can take the examination multiple times if you decide to try to improve your score. The cost of taking the examination varies depending on the country where the examination is taken.

Graduate schools frequently do not require the TOEFL for the following individuals:

- Those who hold academic degrees from postsecondary institutions in English-speaking countries such as Canada, England, Ireland, Australia, and New Zealand
- Those who have completed at least a 2-year course of study in which English was the instruction language
- Those who have been students in a school where English was the language of instruction in an English-speaking country for at least 2 years

If you are a non-native speaker, you should contact the admissions office of the institutions where you plan to apply to determine if you should take the TOEFL. Scores on the TOEFL are usually accepted by schools if the exam was taken within the past 2 years.

■ GRADUATE FELLOWSHIP AWARDS INCREASE YOUR OPTIONS

Many graduate fellowships are available, and a fellowship grant can be a major factor in being accepted into a highly competitive graduate program. Schools evaluate your application from the perspective that you have demonstrated that you are academically competitive (because you got the grant) and you are basically a "free" (in terms of money) addition to their program. Some fellowships are restricted, such as the NCAA Postgraduate Scholarship, which is only given to scholar athletes. Other fellowships such as the NSF Graduate Fellowships are open to all but are very competitive. However, if you do not apply, you have no chance to obtain a fellowship. So, consider applying. NSF Graduate Fellowships provide tuition, living expenses, and even support limited travel to professional meetings. Application to the NSF Fellowship must be initiated early in the fall a year before your anticipated matriculation. Check the NSF website (http://www.nsf.gov/funding/) for deadlines and appropriate forms. Review the application to determine if you have the profile and background required. One major component of the NSF Fellowship is a statement summarizing a potential research project. Do you have enough background and information to write such a research proposal? Your proposed project should exhibit knowledge of current research (literature search), propose to test an appropriate scientific hypothesis, and describe a realistic project that can be completed in the fellowship timeframe. Consult with your faculty advisor and research mentors about writing the research project section. NSF fellowships require several recommendation letters from your professors and research mentors. You can also apply for the NSF Fellowship during your first year of graduate study, and for many students it is easier to write a quality proposal at that time.

There are a relatively large number of fellowships available to graduate students. Many of these have significant restrictions in who can apply. Look at the website https://www.grad.uiuc.edu/fellowship/category/11 from the Graduate College at the University of Illinois at Urbana-Champaign. It has an extensive list of fellowships for science/engineering study as well as for women and underrepresented minorities.

Many graduate fellowships are available for underrepresented groups. Possible programs can be found at the following websites:

1. Fellowship for historically underrepresented minority groups: http://www.aauw.org/education/fga//

2. Graduate fellowships for African-American men and women: www.uncf.org/merck/programs/grad.htm

3. Guide to Graduate and Professional School Fellowships: a list of websites identifying scholarships for minority students: http://www.imdiversity.com/Villages/Channels/grad_school/gs_fellowcontents.asp#supportmed

4. A directory of graduate school fellowships for minority students: http://www.gradschools.com/diversity/fellowships.html

All these fellowships would greatly enhance your application and make you more attractive to a highly selective graduate program. Individual schools also may have minority fellowship programs that are specifically for graduate students at their school.

■ ADMISSIONS ISSUES FOR PhD APPLICANTS WHO RESIDE OUTSIDE OF THE UNITED STATES AND CANADA

Some graduate schools have developed long-term relationships with faculty at foreign academic institutions. The PhD programs may offer summer undergraduate research experiences each summer to several students from the foreign institution. They may invite the faculty to do sabbaticals in their research laboratories. Faculty from the graduate school also may visit the foreign education programs to give research seminars and provide recruitment sessions for the undergraduates. If you are a student at a foreign undergraduate institution, ask your faculty if your school has such arrangements with a U.S. graduate degree program. Alternatively, one or more of your faculty may have a collaborative or a personal relationship with a faculty member at a U.S. graduate school. Your ability to gain admission to the U.S. school may be facilitated by such a relationship.

Foreign student applications present a unique set of challenges for admissions committees. Most of these applicants come from education institutions that are not well known to many American faculty.

In addition, many of the applicants will not be able to appear at the school for a personal interview. Education programs have devised a variety of approaches to evaluate these students. Some programs engage their faculty who were born or educated outside of the United States and Canada in the admissions process. Admissions committees frequently struggle to determine if the GPA from a less-well-known U.S. school is comparable with schools they know. Foreign undergraduate programs can present even a greater challenge. During the initial screen of applicants, the foreign born/educated faculty may help the admissions committee evaluate if the school where the foreign applicant was educated is competitive to U.S. schools.

As we discuss later in this chapter, admissions committee members tend to favor letters of recommendation from faculty they either know directly or at least know of their reputation from their research publications. Foreign faculty who write letters of recommendation may not be known by members of the admissions committee. One way admissions committees deal with this issue is to have their foreign born/educated faculty contact faculty at the school where the student was educated and request that individual to provide an additional reference for the applicant.

■ THE APPLICATION PROCESS FOR MD-PhD PROGRAMS

The centralized application process for both MD and MD-PhD programs is managed by AMCAS. As described later, applications for MD-PhD programs have additional components and may have different deadlines as compared with MD applicants from the same school. Detailed information on the AMCAS application is available at http://www.aamc.org/students/amcas/amcas2007instructions.pdf. The primary application, application fee, and academic transcripts must be received at AMCAS by deadlines set by individual MD-PhD programs. You should apply as early as possible in the application cycle. If at all possible, submit the application close to the opening application date. Schools usually have additional application requirements (supplementary or secondary application materials including recommendation letters) and have different deadlines for these secondary materials, which are requested if you pass the initial screening. Information concerning these program-specific requirements can be found on MD-PhD program websites. It is important that you check application requirements for each program where you plan to apply.

In the online AMCAS application form, sections that are shared by MD and MD-PhD applicants identify the courses you have taken, describe the work and extracurricular activities that have enhanced your intellectual development and leadership skills, and provide a personal statement concerning why you want to attend the medical school. Applicants must request the registrar from each academic institution they have attended to forward an official transcript to AMCAS. AMCAS has a transcript request form that should be provided to registrars when you request transcripts be sent to AMCAS. Another component of the application is your MCAT scores. Some MD-PhD programs allow applicants to just submit their MCAT scores, but it is usually is an advantage to submit both the MCAT and GRE scores.

What are the components of the MCAT?

The MCAT is a standardized, multiple-choice format examination that is designed to assess your problem-solving, critical-thinking, and writing skills. The examination also evaluates your knowledge of science and principles related to the study of medicine. Examination scores are reported in four areas:

1. Verbal reasoning
2. Physical sciences
3. Biological sciences
4. Writing

The exam is scored 0–15 for the verbal reasoning, physical sciences, and biological sciences sections. The writing section is scored H–T (*H*orrible to *T*errific). Information on preparing for the exam, exam content, fees, dates and deadlines, and registering to take the MCAT can be found at www.aamc.org/mcat. The examination is given at multiple test centers throughout the United States and the world on many dates. You should take the examination in the spring during your junior year or take the examination at least 16 months before you expect to matriculate in a MD-PhD program. You should spend several months preparing for this test. There are review books and expensive commercial courses that can be taken to help you prepare for the exam (see Appendix B). Most students can adequately prepare on their own using online practice tests and MCAT preparation books. Your scores on this examination are very important to even be considered for an MD-PhD program. If you do not have scores that meet the degree program's minimum, your application will be rejected when the program evaluates your primary application. Therefore make a great effort to adequately prepare for the exam. You can retake the GRE or MCAT if your scores are not at the level of being competitive. Most schools look at all MCAT or GRE scores of an applicant. Do not take either test without proper preparation just to see how you would do because all scores count. Check with the MD-PhD programs that you are interested in to determine the average scores of accepted students. Generally, any individual exam score below 9 is problematic, and a total score of 33 is needed to be seriously considered by most MD-PhD programs.

Personal statements in MD-PhD applications

The personal statement is your opportunity to describe your unique characteristics that set you apart from others who may be applying to the medical school and why you are interested in a career that includes medicine and research. You need to demonstrate that you have characteristics of integrity, judgment, compassion, intellectual curiosity, motivation, resilience, and dependability. You must document your personal experiences in medically related activities and describe how these experiences affected your desire to become a physician. Outline how your interest in medicine developed and changed over time. Discuss what challenges you have faced and how you met those challenges. Give evidence of your strengths and attributes. Use specific examples or anecdotes if appropriate. Let your personality show in your writing, but do not become so creative that you detract from your strengths. In addition, include a description of your research experiences and how these experiences have shaped your interest in an MD-PhD degree program. It is also important that you align your personal goals with the mission and training goals of the school where you are applying.

If you are a non-science major or a nontraditional student who did not take the classic pre-med course schedule, you should explain how your interests in science and medicine evolved. If you are not a recent college graduate, your MCAT scores may not reflect your academic ability. In this case, it is important for you to have taken science classes during your post-undergraduate period from a school with rigorous academic standards and received high grades in the classes. If there are deficiencies in your academic record, you should explain how you positively addressed the problems and demonstrated the tenacity to overcome adversity (turn a potential negative into a positive). As you prepare to write this essay, it may be of value to look the AAMC website (http://www.aamc.org/data/msq/start.htm) that shows responses first-year medical students gave to a variety of questions concerning their pre-med experience, the medical school application process, reasons for selecting a medical school, and future career goals.

Letters of recommendation for MD-PhD applications

Once you have passed the initial screening of a school's MD-PhD program, you must request letters of recommendation from your pre-medical advisor, other professors, and your research mentors as components of your secondary application. Most colleges have a health professions (pre-med) committee that prepares a committee letter of recommendation. Check with pre-med advisors early in the process, at least several months before submitting your AMCAS application, to request the committee letter. Determine what information the committee needs and if an interview is required. Most committees start the process, including the interview, in the spring of your junior year, and the committee letter is prepared before you even initiate the AMCAS application. Failure to obtain recommendation letters from your research mentors or your pre-med committee in a timely manner will make your application uncompetitive. The individuals who write letters for you should be familiar with your academic abilities, research experiences, career goals, and the reasons you desire a combined degree program. The information previously discussed for letters of recommendation in the section for PhD programs is also applicable to MD-PhD applications.

Essays specific to MD-PhD programs

MD-PhD applicants write two additional essays for the AMCAS application. The first concerns why you want to pursue the combined program, and the second describes your significant research experiences. Each of the essays in the AMCAS application form are very important components in admissions committee decisions, and they should be given a great deal of thought. Although you must write the essays, it is a good to discuss your ideas for the essays with your pre-medical advisor and research mentors. Have other individuals read your essay and give you feedback. Do the essays reflect who you are and why you want to pursue an MD-PhD degree? Make sure the essays are free of spelling errors and grammatical mistakes. Many of the issues we discussed previously in relation to the personal statement for the PhD application are applicable to the two MD-PhD essays.

Once you have submitted your application and the requested materials, you should check with each school to make sure your file is complete. Many applicants are at a significant disadvantage simply because some component (such as a letter of recommendation) of their application is missing or late.

■ ADMISSIONS COMMITTEE REVIEW PROCESS FOR PhD APPLICANTS

Once PhD or MD-PhD programs have received their pool of applicants, what is the process by which the programs select applicants to interview? For PhD programs, the admissions committee (or admissions subcommittee) reviews all applications. Most schools are on an internal deadline, not a rolling admission process followed by most medical schools. Many admissions committees provide their members with forms (Box 6-4) on which they summarize information on each applicant to distill the important details of the application. Each admissions committee member completes the form based on the information provided in the student's application. Committee members rate each student, and those with the most positive rating are invited for an interview. The number of applicants invited to interview is determined by the size of the graduate program, available spots for new students, and the amount of funds available to pay some or all of the expenses of interviewing students.

■ ADMISSIONS COMMITTEE REVIEW OF MD-PhD APPLICATIONS

The admissions review process is somewhat different for MD-PhD applicant evaluations. In many cases, the program director makes a preliminary review of each application for its MCAT scores, overall GPA, the science GPA, the breadth of the science classes, and

Box 6-4 Hypothetical Application Summary for Graduate School Admissions Committee Members

Applicant Evaluation Form

Name:

Undergraduate School:

Score: 1 = highest (best), 5 = lowest (worst)

—— **Research experience:** Evaluate the significance, quantity, and quality of the applicant's research experiences. Is there a demonstrated passion for research?

—— **Personal statement:** Evaluate the applicant's academic development, alignment of goals with our program's focus, and understanding of the realities of a research career.

—— **GPA** ——

—— **GRE** (%V = ——, %Q = ——, %A =——)

—— **Advanced exam** (Biochem/Biology/...)

—— **Course quality and breadth of science classes**

—— **School standing**

—— **Letters of recommendation:** Evaluate the important strengths and/or limitations identified in the letters. Include comments on maturity, creativity, curiosity, drive, energy, enthusiasm, commitment, intellectual contribution to the research project(s), organizational skills, and communication skills. Assess the experience of the letter writer.

For MD-PhD applicants:

—— **Clinical medicine:** Evaluate the quality and quantity of medically related activities that the applicant has engaged in.

—— **Leadership/teamwork:** Evaluate how the applicant has demonstrated leadership skills while working effectively as part of a team in both research and clinical medicine.

—— **Career goals:** Does the applicant understand the rigors and nature of a career that combines research and medicine?

—— **Average score**

General Assessment:

Rating:

Top 5%	Top 10%	Top 20%	Top 30%	Top 50%	Top 75%	Reject
1.0	1.5	2.0	2.5	3.0	3.5	4.0

the quality of research experiences to make a selection of the most highly ranked applicants. Usually, the director also takes into account the academic rigor of the applicant's undergraduate institution. Many directors have been involved for years in the selection process and know how competitive the education programs are at different undergraduate schools as well as the quality of the students who have previously matriculated into their MD-PhD program from specific schools. MCAT scores play an important discriminator in this first cut. In some schools, a total of 33 or above points in the biological science, physical science, and verbal MCAT sections is the minimum score, whereas more competitive MD-PhD programs have higher minimum scores. The writing score usually is not a major factor in admission decisions unless the verbal score is low in comparison with the other section scores. Nationally, AMCAS has reported that the average MD-PhD applicant MCAT score is 33.6 (http://www.aamc.org/research/dbr/mdphd/Garrison.pdf).

Applicants who pass this first evaluation are then forwarded to either the MD-PhD admissions committee or, in some schools, the medical school admissions committee for further review. In some MD-PhD programs the selection of applicants is controlled exclusively by the MD-PhD committee, whereas in other programs the applicant must be reviewed by both the MD-PhD program and the medical school. The admissions committee(s) carefully reviews each application in detail and then the committee members rank the applicants. During this initial review, the medical school admissions committee focuses on academic ability (GPA, MCAT scores), substantive medically related experiences, and the variety of additional issues previously reviewed in the discussion of letters of recommendation. The MD-PhD committee looks for a passion for research (depth of research experiences), academic ability, and information in the letters of recommendation. From this list, a group of the highest ranked individuals are invited for an interview. The number of applicants selected to interview varies with the size of the MD-PhD program but frequently is 5 to 10 times the number of students expected to matriculate. Data collected by AAMC suggest that 62% of applicants to MD-PhD programs are invited to at least one school for an interview (http://www.aamc.org/research/dbr/mdphd/Garrison.pdf). Some applicants will not be successful in being highly ranked in the MD-PhD review process, but their application may be considered for the MD or PhD program.

■ SUMMARY

1. Applications for PhD programs usually consist of an application fee, an official transcript of courses from each academic institution attended, GRE scores, multiple letters of recommendation, and a personal statement that includes a description of your involvement in research and your education goals.

2. The GRE currently consists of verbal reasoning, quantitative reasoning, critical thinking, and analytical writing sections. It is very important that students invest adequate study time to prepare for these exams.

3. The TOFEL is a prerequisite for admission to most graduate programs in the United States for individuals for whom English is not their native language. It is designed to evaluate how well you read, hear, speak, and write English.

4. You should foster interactions with your science faculty and research mentors so they can adequately describe your intellectual abilities, creativity, lab skills, leadership roles, and motivation for a career in research when they write letters of recommendation.

5. Students should be knowledgeable about the issues addressed in a positive letter of recommendation from their faculty and use the information to critically evaluate how they perform in their education/research programs and interact with faculty advisors, research mentors, and laboratory coworkers.

6. The personal statement in your application should describe how you developed your interest in research, your education goals, what motivates you to become a scientist, why you are applying to the school, and your research experiences.

7. It is worth the effort to apply for graduate fellowships, because a fellowship grant can be a major factor in being accepted into a highly competitive graduate program.

8. Most MD-PhD primary applications consist of an application form, application fee, MCAT scores, and academic transcripts that are submitted to the AMCAS. Additional application requirements (supplementary or secondary application materials, including recommendation letters) are subsequently required if you pass the initial screening process.

9. The MCAT evaluates a student's knowledge of science and principles related to the study of medicine. The exam includes verbal reasoning, physical sciences, biological sciences, and writing sections.

10. Most students can adequately prepare for MCATs on their own using online practice tests and MCAT preparation books. Your scores on this exam are very important to even be considered for an MD-PhD program, so make a great effort to prepare for the exam.

11. Both PhD and MD-PhD review committees evaluate each application for its GRE/MCAT scores, overall GPA, the science GPA, the breadth of the science classes, letters of recommendation, and quality of research experiences to make a selection of the most highly ranked applicants to invite for interviews. MD-PhD applications are also evaluated for medically related experiences.

12. In many cases the program director makes a preliminary review of each application for its MCAT scores, overall GPA, the science GPA, the breadth of the science classes, and the quality of research experiences to make a selection of the most highly ranked applicants.

13. Nationally, AMCAS has reported that the average MD-PhD applicant MCAT score is 33.6.

■ REFERENCES

1. Kuncel, N.R., Hezlett, S.A. Standardized tests predict graduate student success. Science 315:1088–1089. 2007.

7

Getting Past the Paper Applications: The Interview Process

Now that you have submitted your applications to PhD or MD-PhD programs, you must wait for many weeks and often months with great anticipation to see how many programs invite you for an interview. Use good judgment in recording a message on your answering machine or using a "creative" e-mail address. Any negative or inappropriate message accessed by the graduate or medical school may be a factor in your admission. If you are going to be out of the country for any significant time during the interview period, it is wise to send a letter to each school informing them of the time period you will be unavailable. In most instances students should not participate in a fall or winter abroad semester when they are in the application process because most interviews are scheduled for the fall or winter.

■ SCHEDULING INTERVIEW VISITS

Once you receive invitations, most programs have several dates that you can select for a visit to the school. You should resolve potential conflicts that you may have with your college classes or personal commitments and schedule your visit promptly with each program. It is best if you visit the schools during the dates they offer you because tours and other organized events are included. Many programs go out of their way to arrange a special interview date for you if you are unable to attend one of the prescheduled dates. However, you usually gain more information on the PhD or MD-PhD program and the faculty and students in the program if you interview on the program's prearranged dates. Carefully follow the instructions provided by the graduate school or MD-PhD program concerning arrangements for airline and hotel reservations.

■ HOW TO DEAL WITH TRAVEL PROBLEMS

Schools are accustomed to having bad weather interfere with applicants' travel. If this happens, phone the program's office and work out an alternative interview schedule. Similarly, if you encounter other problems traveling to the interview, talk to the staff in the admissions office. They will likely do everything they can to help you solve your travel problems. If you fly to your interview, take your luggage as a carry-on rather than checking your luggage with the airline. The odds of the airline losing your luggage are directly proportional with the importance of this interview to your future education.

■ DO YOU PAY FOR YOUR INTERVIEW TRAVEL EXPENSES?

There are significant differences in the amount of reimbursement you may receive for travel expenses from PhD and MD-PhD program interviews. Some programs pay almost all expenses, whereas others may only provide meals during your visit. Many programs offer you the opportunity of staying with a graduate student or a MD-PhD student, which may be very insightful. If expenses are covered by the graduate school, you should not hesitate to ask the staff of the admissions office how they reimburse your expenses. The interview is a very important component of the selection process, and you should go to as many different programs as you receive offers. If necessary, borrow money to participate in these interviews. Money spent to interview may save you much more in the future if the program doesn't serve your needs. Your visit to a school may dramatically change your opinion of the school's education program. You may find that you really like a school where, for a variety of reasons, you may have been reluctant to apply. Alternatively, a school that was your first choice when you submitted your applications may not meet your expectations when you actually visit the campus. Your personal interactions with the faculty give you the opportunity to impress them that you are an exceptional candidate for their program. Interviews can have a very positive effect on your chances of being accepted to a PhD or MD-PhD program.

■ DRESSING FOR INTERVIEW DAY

On the day of the interview, arrive on schedule and dress appropriately. Interviewers may quickly judge interviewees based on their appearance, so have your hair clean and trim, remove all visible body piercings, and appear neat for your interviews. Faculty interviewers of PhD and MD-PhD applicants have different expectations of what clothing is appropriate for both them and their students. Graduate school faculty frequently dress informally and do not expect applicants to wear business suits or dresses. In contrast, physicians who see patients frequently wear business apparel and may expect applicants to dress similarly. If you are in doubt about what to wear, phone the degree program office and ask the program's administrator about what applicants typically wear to interviews. Do *not* bring family members, spouses, or friends to an interview. This is show time for you and only you. Get plenty of sleep before your interview.

■ WHAT HAPPENS DURING A TYPICAL INTERVIEW DAY?

We first discuss the components of the interview process that are common to both the PhD and MD-PhD programs and then discuss aspects that are unique to the MD-PhD program. The structure of all PhD and MD-PhD program interviews is designed to both evaluate the applicant and recruit applicants to their program. Although it may not be obvious to you, the interview schedule provides time to attain both goals. Most interview days begin with an introductory session explaining how the school's curriculum is structured; the stipend funding mechanisms; an overview of the research, education, and clinical facilities; and a description of the community environment. There will be one-on-one (or possibly group) interviews with the program director and multiple members of the admissions committee. The faculty interviewers receive a copy of your application (or summary of your application), but they may not have time to read the entire application. Therefore do not be offended if you are asked to summarize the information in your application. In some programs you have the opportunity to select additional interviews with faculty whose research interests you. Take advantage of meeting with the faculty because you can learn about laboratories where you potentially could do your thesis research. The faculty who participate in these interviews also may become advocates for your admission. Some programs also include an interview with PhD or MD-PhD students. Time is usually set aside for you to tour the classrooms and research and technical support (core) facilities so you can observe directly the environment where you might do laboratory research and take classes. Although some of these tours may be optional, such tours can be very informative. You should participate in all activities associated with the interview program. You can catch up on your sleep on the way home. Most programs also have either meetings or social periods with current PhD or MD-PhD students where you can get a student's perspective of the school's educational environment (discussed later). The interaction with students is very important and allows you to get insight into the graduate school culture and the school's academic environment. Do not ask graduate students inappropriate questions or make negative comments about faculty who you meet during the interview day. The graduate students may be contacted by the faculty concerning the applicant and asked for their impression of the individual.

■ HOW TO PREPARE FOR FACULTY INTERVIEWS

Preparing in advance for each faculty interview is one of the most important things you can do to enhance your changes of being selected to matriculate into either a PhD or MD-PhD program. The interview is one component of the application process that many students neglect. *The fastest way to negate your productivity resulting from 3 or more years of hard work in your undergraduate education is to not adequately prepare for a 30-minute interview!* Most importantly, you must be able to clearly articulate that you have a passion for science and, if a MD-PhD applicant, for medicine as well as science. This passion should be apparent when you discuss your research experiences. You also need to demonstrate from your research experiences that you understand the positive and negative aspects of a research career. Before going to the interview, use the notes you prepared during each research experience to review your lab projects. Also, reread the review articles you collected during your research to help remind you of your project's background. Reread your personnel statement and essays because many questions may evolve out of them. This is your opportunity to articulate your academic and personal strengths that make you a strong applicant for the school's degree program.

Think about what you can add to the review process that is not already on your application. What do you want the interviewer to know about you? Although you do not want to come across as boastful, you want to clearly highlight your strengths during the interview. A useful website for some students interested in medical schools is found at a commercial site (www.interviewfeedback.com). Be very careful what information you add to the site as many medical schools monitor the site. Also be very careful what pictures or comments you post on sites such as MySpace and Facebook because faculty have been known to surf the sites and an inappropriate picture or comments can be damaging. Do not make the mistakes (discussed in Box 7-1) that some past interviewees have made during their interviews with graduate school faculty. Think about how you might respond to questions from interviewers (see Box 7-2 for a more extensive list of potential questions you may encounter).

When you discuss each of your research experiences, you should be able to:

- Clearly and concisely describe both the theoretical and experimental aspects of your research projects. What did you learn from your experience?
- Identify your creative contribution to each project.
- Describe the techniques that you used in your research.
- Discuss both the positive and negative aspects of your projects so you convey that you were able to appropriately deal with setbacks.
- Use the description of your research experiences to show you are highly motivated and have a strong work ethic.
- Be prepared to discuss how you worked and interacted with other individuals in the lab.

Box 7-1 Which of These Applicants Will Receive Positive Evaluations from Their Interviewers?

- Applicant 1 has a high grade point average from a large university but interjects "like" and "ya know" in her sentences at least 25 times during the first 15 minutes of the interview conversation.
- Applicant 2 has prepared a written list of two questions about the research of each faculty member she will meet during her interviews.
- Applicant 3 either looks out the window or at the wall of pictures in the room during most of his interview with the program director.
- Applicant 4 is asked to describe the most important issues he will consider when selecting a graduate program. He responds by saying he is interested in the level of the program's stipend and the number of classes he is required to take to meet his degree requirements.
- Applicant 5 has several glasses of wine during the social hour and then proceeds to go through his repertoire of sexually explicit jokes with the graduate students in the program.
- Applicant 6 brings her mother along to the social events during interview weekend.
- Applicant 7 (male) asks one of the male graduate students at a social event which of four female graduate students are "available and like to party."
- Applicant 8 declines to attend several of the optional tours of the institution's research facilities during interview weekend because she is "tired."
- Applicant 9 encounters a weather delay in her flight, arrives at her hotel at 2 a.m. the day of her interviews, and is without her luggage because it was lost by the airline. She gets up at 6:30 a.m. so she can meet with the program director 30 minutes before the start of her interviews to explain why she is dressed in jeans and a sweatshirt for her MD-PhD interviews.
- Applicant 10 has two difficult faculty interviews in the morning session. After lunch with the other applicants, she assails the program director's administrator about the quality of the food.
- Applicant 11 notices the photos on the interviewer's computer screensaver and spends half of the interview exchanging stories about fly fishing in Rocky Mountain lakes and streams.

Box 7-2 Prepare Responses to These Interview Questions

A. RESEARCH EXPERIENCES

1. Tell me about the hypothesis you addressed in your last research experience.

2. Tell me about the hypothesis you addressed in your other laboratory experiences.

3. How did you interact with your mentor to learn about the process of doing laboratory research?

4. What are your strengths and weaknesses as a laboratory investigator?

5. Tell me about your worst laboratory experience and how you addressed the problem.

6. How did you contribute to the research environment of your laboratory?

7. What work schedule did you keep in the laboratory?

8. What was your intellectual contribution to the research project?

9. Describe how you identified previous research (background) that was done on your project.

10. What was the best part of your research experience?

11. How do you deal with the stress of having a series of failed experiments?

12. Did your undergraduate education give you the science background you needed to work on your laboratory project?

13. What laboratory skills or abilities would have made your research project more productive?

14. Describe a laboratory technique that was difficult for you to learn.

15. Describe the types of oral and written presentations you made concerning your laboratory project. Were the presentations stressful? Were they constructive learning experiences?

16. What motivates you to go back to the lab after your experiments have failed?

17. Were there individuals in your laboratory with whom you had difficulty interacting personally?

continues

Box 7-2 continued

18. How did you decide that research was a career that you wanted to pursue?
19. Tell me about a laboratory situation in which you showed initiative.
20. Tell me about a laboratory situation where you demonstrated leadership.
21. Tell me about a mistake you made in the laboratory and how you handled it.
22. What aspects of laboratory research are "fun"?

B. PERSONAL CHARACTERISTICS

1. What is your strongest personal asset?
2. What are your strengths and weaknesses as a student?
3. What issues contributed to your decision to attend your undergraduate school?
4. What courses did you enjoy the most?
5. Which courses were most difficult for you?
6. What experiences during your undergraduate education gave you the most satisfaction?
7. Do you believe you got a good education at your undergraduate institution?
8. During your undergraduate education have you worked to your full potential?
9. What were the shortcomings of your undergraduate institution?
10. Do you believe your academic record or standardized test scores accurately reflect your abilities and potential?
11. What academic or personal setbacks have you encountered in your life? How have your responses to these setbacks changed you as an individual?
12. What will you do if you are not accepted into one of the PhD or MD-PhD program to which you have applied?
13. What are the major issues you will consider when you select an advanced degree program to attend?
14. What other degree programs have you applied to besides our program?
15. What are your long-term career goals?

16. How will our degree program help you achieve your career goals?

17. Describe the process you used to evaluate different career options in science.

18. Describe a recent experience that has been difficult for you and how you subsequently used introspection to reevaluate your personal motivations or attitudes.

19. Tell me about a time when you encountered a confrontational situation with another individual. How did you handle it to resolve the conflict?

20. Do you have any personal issues that might keep you from completing your degree program?

21. What do you believe are the strengths and weaknesses of our degree program concerning training our students for professional careers in science?

22. Describe how your interests in science align with the interests of our program's faculty.

C. EXTRACURRICULAR ACTIVITIES

1. What activities do you enjoy most outside of the classroom?

2. What extracurricular activity has given you the most personal satisfaction?

3. Describe recent volunteer activities in which you have participated and identify the contributions that you made to the goals of the activity.

4. Describe how an extracurricular activity has changed your outlook on life.

5. Describe an extracurricular activity in which you have demonstrated leadership skills.

6. Tell me about how you achieved a major accomplishment in your life.

7. Tell me about a recent mistake you made and how you handled it.

8. Tell me about a time you were faced with a difficult situation and how you handled it.

9. What community-related activities are important to your lifestyle (exercise or sports facilities, music groups, bike trails, restaurants, etc.)?

- Respond with clear, concise overviews of your research. Do not ramble on. Interviewers expect students to talk passionately about their research.
- Keep your research project in perspective; don't overinflate the importance of your results or contributions.

You should know the research strengths of the institution where you are interviewing and be able to describe how your interests in science align with the interests of the program's faculty. This information concerning faculty research interests should be available on the program's website. Also review the mission statement of the institution.

You must be able to clearly articulate your educational goals as well as your passion for science and, if a MD-PhD applicant, for medicine as well as science:

- Why do you want to participate in a PhD or MD-PhD program? For MD-PhD applicants, be able to articulate why you selected a MD-PhD program rather than a PhD or MD degree program.
- What is important to you in selecting a graduate school or medical school (class options, collaboration, faculty research interests, opportunities to develop communication skills, collegial environment, competitive investigators who publish in high-impact journals, productive graduate students who publish in high-impact journals, supportive research mentors, well-equipped research labs and core facilities, stable funding environment, faculty committed to graduate education, strong clinical training program)?
- Describe how the particular PhD or MD-PhD's education program (classes and research) where you are interviewing will help you meet your educational goals to become a competitive investigator or for an alternative professional career in science. It is very important that you prepare for this discussion by reading the program's literature describing the faculty's research interests and the academic program in advance of your visit. Do your homework!
- Outline your long-term career goals (acknowledge that this may change with additional education and experience).
- Prepare answers for the questions outlined in Box 7-2. You may hear these questions again at your interview.

Write out your responses to each of these issues and practice discussing each with one of your faculty. In addition, you should be prepared to describe both the positive and negative aspects of your academic undergraduate education to demonstrate that you are an intellectually mature individual. Although your academic transcripts demonstrate that you have a strong academic background, you want to verbally emphasize that you are an independent learner rather than a student who is primarily interested in obtaining good grades. You may be asked to give evidence of your aspirations and learning style. Do not ask about vacations, leave policies, or deferments.

■ BE PREPARED TO DISCUSS THE POTENTIALLY NEGATIVE ISSUES IN YOUR APPLICATION

If you had low grades from a limited period of time during your undergraduate education, you may be asked to explain the reason for the lapse in your academic performance. Don't volunteer information or an explanation about a low grade or test but be prepared to give a response if asked. Everyone has downturns in their life. You need to explain how these temporary setbacks were overcome and what you learned from the setback. Don't give the appearance of being a victim or blame the professor. Don't overexplain. Similarly, if you received low scores on one or more components of the Graduate Record Examination (GRE) or Medical College Admissions Test (MCAT), you may be asked to explain the low scores. Be prepared to give an honest, positive answer. Some liberal arts colleges use examinations that rely on essay-type formats rather than the multiple-choice format in the GRE and MCAT. Students from such colleges may underperform on the standardized exams. Illness may have affected your test performance, but this can sound like a lame excuse. Don't say that you do poorly on standardized exams, because if you are applying to an MD-PhD program, the U.S. Medical Licensure Examination (USMLE) Step One, Step Two, and Step Three exams are also in a standardized format. It is better to emphasize your strengths; suggest your grade point average and productivity in your research experiences are a better reflection of your academic abilities. As previously discussed, some faculty do not believe GRE scores are a useful predictor of future research productivity in graduate school. Medical school admissions committees tend to be less forgiving when it comes to low MCAT score, especially scores in any exam section below 9. Many medical school faculty believe that MCAT scores are a fairly accurate prediction of a student's performance on the USMLE Step One, Step Two, and Three board exams. Medical schools publish the average MCAT score of the students in their matriculating class. This provides pressure to not recruit students with low scores even if other components of their application are outstanding.

■ PRACTICE YOUR INTERVIEWING SKILLS

Many colleges help you prepare with mock interviews and provide the opportunity to videotape a practice interview session. Spend time with your science faculty or staff in your school's career counseling office practicing responses to questions that you might be asked during an interview. Carefully consider what additional information you want to work into the interview that interviewers do not already know about you. Interviews with graduate school and medical school faculty can be intimidating even if you are an extrovert. If you are an introvert, you need to work even harder to develop your interviewing skills. Some faculty may ask you questions that attempt to get you away from your prepared discussion of your strengths. Answer the questions, but steer the conversation back to highlighting your strengths. Some faculty use a situational interview approach that is more difficult to prepare for. This requires you to describe positive or negative events in your research or personal life and describe how you constructively dealt with the event. (See references in Appendix B for sources describing how to deal with situational interviews.)

■ QUESTIONS TO ASK RESEARCH FACULTY

In advance of your interview trip, request a list of faculty with whom you will interview. Prepare for the interview by identifying the research interests of each of the graduate school faculty. This information is usually on the department's website. Prepare written questions to ask each interviewer about their research. These questions do not have to require an in-depth understanding of the investigator's project but could focus on potential applications of the project, how many students are involved, or what techniques are used. Most investigators love to talk about their passion—their research. It is okay if you do not understand the technical aspects of their research. This discussion helps to make the interview bidirectional rather than just focused on you.

Also prepare questions to ask each interviewer about different aspects of the school's education program (lab rotations, types of required versus optional classes, structure of comprehensive exams, collaboration between faculty or between students and faculty in different labs, structure and role of the thesis committee what will guide a student's research thesis). MD-PhD applicants could develop questions regarding research within the program, about the medical school curriculum, and the approach used by the school in educating their students. Is the curriculum a lecture format or is it case study based (problem based)? Ask questions about how the MD-PhD students are integrated into the medical school program and how the program is structured to assist students to balance their research, basic science class, and clinical instruction workload in the dual degree program. How many years do most students need to finish the MD-PhD? Where are their graduates now? What is the average number of years students are in graduate school? What percent of entering students do not finish the MD-PhD or the PhD degree? Structure questions to solicit information rather than being critical or negative. This discussion will potentially enhance your understanding of the details of the education program as well as demonstrate your interest in the program. Anticipate that you will have periods of silence in at least one interview. Either the interviewer has run out of questions or he or she is trying to observe how you respond to a potentially uncomfortable situation. Relax and be comfortable with silence—the interviewer will eventually say something.

■ BASIC INTERVIEWING SKILLS

- Present yourself as animated, energetic, and motivated.
- Shake hands as you introduce yourself to the interviewer.
- Maintain eye contact throughout the interview.
- Do not chew gum.
- Do not look bored or tired (even though you maybe worn out after six or more interviews). Watch your body language.

- Carefully listen to the interviewer's questions and do not interrupt the interviewer.
- Think before answering the interviewer's question.
- Clearly answer interviewer's questions but do not ramble on about unrelated issues.
- Show interest in the school's education program—ask the interviewers what they believe are the strength and weaknesses of the program.
- You may be asked to identify and comparatively evaluate the other programs where you have applied. Do not be critical of other programs where you have interviewed. Emphasize the positive aspects of the program where you are interviewing. Do not rank the institutions even if you are asked to do so.
- Do not present yourself as a "gunner" or an ultra-competitive individual. One admission director of a prestigious eastern medical school recently stated that arrogance during an interview is the most common reason students are not selected by the medical school after the interview process.
- Shake hands (firmly) with the interviewer at the end of the session, make eye contact, and thank the interviewer for their time and willingness to answer your questions.

■ ISSUES YOU SHOULD ASK ABOUT DURING YOUR INTERVIEW VISIT

We previously discussed in Chapter 5 some of the issues you should consider when trying to identify advanced degree programs that will help you meet your education goals. As you prepare for your interview day, it is a good idea to reread the issues discussed in Chapter 5. These issues included the flexibility of the curriculum, faculty productivity (publications, grant support, interprogram research collaborations), and student productivity (publications, post-degree placement). Most interview days have sessions that describe the program's curriculum. You should focus on determining if the curriculum is "one size fits all" or if the curriculum is flexible to enable you to meet your education goals. For example, if you want to focus in neuroscience or immunology, does the curriculum offer classes in that area of science?

How are student stipends funded?

In many programs your stipend is funded by the education program for the first 1 or 2 years. During the research phase of a student's thesis, your stipend may be funded by the thesis advisor's research grants. What happens if the thesis advisor loses his or her research grant funding? Will the education program pick up your stipend or will you be forced to find another research lab to continue your thesis research? The latter option might entail starting another thesis project. You should ask about the number of faculty in the education program who do not have extramural research funding. Besides potentially not having funding for student stipends, faculty without grants may have serious problems funding their research laboratories (supplies, equipment, and technician salaries).

Graduate student productivity

Information concerning student productivity can be difficult to obtain unless the information is posted on the education program's website. However, once you are on campus for your interview, such information should be available from the admissions committee staff. It is frequently prepared for the institution's grant applications. Ask the staff for a recent list of student publications. We discussed in Chapter 5 how to evaluate the quality of the science journals where the students have published. Another way to evaluate student productivity is to identify where recent PhD graduates have gone after they have completed their PhD or MD-PhD degrees. Have PhD graduates gone on to postdoctoral research or research-intensive residency programs? Have PhDs accepted positions in high-quality academic institutions, taken jobs in industry, taken positions as undergraduate college instructors, or gone into non-research–related careers? Have the graduates moved on to other degree programs (e.g., MBA, JD, MPH [discussed in Chapter 2])? Have MD-PhD graduates gone to research-intensive clinical fellowship programs? Which of these positions are consistent with your education and career goals? PhD and MD-PhD programs have to compile this information for National Institutes of Health (NIH) training grant applications, so it should be readily available from the program. As with publications, graduate programs that are successful in educating future research investigators tout their success in getting their students into high-quality research postdoctoral fellowships. Use this information to determine if the outcome of recent graduates aligns with your potential education and career interests.

Career options

In Chapter 2 we discussed the variety of career options that can be pursued by individuals with PhD or combined degree programs. Currently, there is a demand in our society for individuals who have a PhD plus additional advanced training in another area of specialization (e.g., law, business, biotechnology, healthcare policy, etc.). There is a transition occurring in many PhD and MD-PhD programs concerning the faculty's expectations for appropriate career options for their graduates. Historically, many graduate school faculty believed that they were only preparing their students for careers in academic research. Being a competitive researcher provided the greatest breadth of career options, had the greatest impact on science and society, and was the goal that their students should seek to attain. Graduates of their program who chose alternative career paths were considered by some faculty as "program failures." In part, this attitude exists because graduate school and some medical school faculty must successfully compete for research grants and must regularly publish in high-impact science journals to maintain their academic position (and their self-image). The economy today provides many career opportunities for PhDs that did not exist 15 to 20 years ago. These careers can be intellectually, socially, and financially rewarding. A few PhD programs currently offer combined degree options to their students. Thus you should ask the faculty and admissions staff of the school where you are interviewing if their recent graduates have chosen these alternative career options. You should consider selection of alternative career options as a strength of the program because it may provide you with additional career options once you have completed your degree. However, if you ask about alternative career training options in the graduate school, direct your question to the degree program's administrator rather than to the research faculty.

Time to degree

Another question that can help you evaluate student productivity and the organization of the degree program is the average number of years students take to complete their degree program. If the average time to graduation is greater than 6 years for the PhD program or more than 7 for the MD-PhD program, suspect a dysfunctional program where student thesis research is not appropriately mentored. There are valid reasons why some students take longer to complete their degree. However, most students should complete their degrees in a timely manner if the faculty are providing appropriate guidance to their students.

Identify potential teaching requirements in the program

In many graduate programs, the faculty teach undergraduates, nursing, allied health, and/or medical students. The faculty may use graduate students as teaching assistants to present lectures, organize and supervise laboratory sessions, grade exams, tutor students, and lead course reviews. Participating as a teaching assistant can be a valuable educational experience for graduate students. Teaching skills are important to learn even if you plan a career other than research in industry or government. However, being required to commit a large component of time as a teaching assistant can detract from your education and research objectives as a graduate student. If information on teaching requirements is not described on the graduate school's website, be certain you find out about this when you interview at the school. Ask about it in a positive manner indicating potential interest in developing your communication skills.

Identify how students are trained to think on their feet

Graduate education is more than the sum of classes offered. Students need to learn verbal presentation skills. This includes making coherent journal club presentations summarizing a recently published journal article or giving a presentation on your recent research accomplishments. For most of us, the skills needed to teach others in formal or informal presentations or to convince others about how your research data support your conclusions are learned through (sometimes painful) experience. You must learn how to effectively think on your feet and make effective presentations. Because these skills are learned through repetition, the graduate program must incorporate them at frequent intervals during a student's training program. You should identify the type and frequency of required student presentations in the graduate program. You should also ask the current graduate students if the faculty provide a constructive, supportive environment that helps students develop their presentation skills. You do not want to participate in a graduate program where the faculty routinely denigrate their students in seminars, journal club presentations, or in classes.

Scientific interactions with research faculty

Science is most productive (and fun) when it occurs in a collaborative environment, so ask how easy it is to interact with individuals in other research laboratories. Are there formal interactions (multiple lab research meetings) or informal mechanisms for sharing ideas and reagents within the department or program? Do the labs have regular meetings to discuss research plans, progress, and techniques? Are the research faculty available for discussions? Does the research advisor participate in the lab as a bench scientist where they have the opportunity to interact directly with students? Alternatively, does he or she travel frequently, thereby limiting mentoring time with students? Some of the best interactions you may have with your research mentor can occur when you work side by side at a laboratory bench. It is important that graduate students learn about research that is ongoing in other institutions both to be informed about their area of specialization and to develop breadth in their knowledge base. One way to learn about such activities is if the department has a formal seminar program where investigators from outside of the institution are invited to lecture. How often do outside speakers visit the program and do students have the opportunity to interact with visiting faculty? Are graduate students encouraged to attend professional meetings where they can present their research results, meet other scientists in their field, and are given financial support for travel? You should ask about the computer support facilities, especially if all students are required to have laptop computers for their classes.

What is the community like where you are going to live?

You are going to live in the community during the 5- (or more) year period you are working on your degree. Ask about the quality of life outside of the classroom and laboratory. Does the community provide activities that are important to your lifestyle (exercise or sports facilities, music groups, bike trails, restaurants, etc.)? Are there frequent social interactions between students and other members of the community who may share an ethnic background or outside interests with you? Will you have access to quality, affordable housing within a reasonable distance from the school? Is parking available to students? Are there personal safety issues on campus or in the community? Will you and your family have access to affordable health insurance and medical care? Are there good support and a network for spouses and families?

■ HOW TO ADDRESS THE NEGATIVE ASPECTS OF INTERVIEWS

Although we have emphasized many aspects of your interview you can control, there also are illegal or inappropriate questions that interviewers may ask you. These issues include your age, marital status, height, weight, financial resources , religion, race, sexual issues (pregnancy, gender preference, family planning, marriage plans), physical and mental disabilities, drug use, or personal family issues. Politely decline to respond to such questions by indicating the answer to the question will not affect your performance as a medical student or graduate student. However, in your response you do not want to cross the line between being firm and attacking the interviewer. It is possible the interviewer did not intend to make an inappropriate comment and will return to a civil discussion. Do not let the interview go up in flames unnecessarily. It is permissible for an interviewer to ask you a question, such as "Are there current or foreseeable family obligations or health considerations that would prevent you from being at school or work every day or that would make it difficult to work whenever you are needed?"

Some interviewers are intentionally aggressive and even hostile. You may not like this approach, but it is not illegal. Do not permit yourself to be baited into making an inappropriate response. If you introduce a subject such as being married or a personal health issue, then the subject can be addressed by the interviewer. Take your time and answer the questions in a calm and professional manner. If you believe an interviewer has made inappropriate comments, you should speak to the program director while you are at the institution in an attempt to eliminate the comments of the interviewer from your application evaluation. Program directors cannot control what their faculty discuss during interviews. Program directors will have eliminated faculty as interviewers if they have previously demonstrated unprofessional behavior, but new interviewers can present new challenges to program directors. It is likely the program director will be sympathetic and respond appropriately. If this does not work, seek out advice from your faculty advisor about your interview and how to handle your concern. You could write a letter to the dean of the school after you have returned home. Taking this action may effectively eliminate your chance of being accepted to the school. However, do you really want to spend 5 or more years of your life dealing with faculty who treat their students shabbily? You have the option of taking legal action against the school, but such lawsuits are expensive, time consuming, and difficult to prove.

■ THE POTENTIAL TRAPS OF SOCIAL EVENTS

Most PhD and MD-PhD programs include a social period where applicants interact with faculty and/or students. You want to keep in mind that regardless of what happened during your interview day, you must maintain a mature, positive attitude with everyone (administrative staff, faculty, students) your encounter during your conversations. Do not ask inappropriate questions or comment on other institutions you are currently interested in attending, especially with other applicants. Do not name drop or gossip about other institutions. Be careful! Some faculty use these social events to learn about the applicant's personality when they are in an informal setting. Dress appropriately and act professionally. This is not the time for you to relax—maintain your best manners. You want to be in control of your behavior, so do not drink alcohol during your visit. Engage the students in friendly conversation and try to identify common interests but do not make negative or inappropriate comments (i.e., jokes, sexually or racially oriented comments). Remember that students may report comments and actions that you make to members of the admissions committee. Because many students are candid in their conversations with visitors, this can be a valuable time for you to learn about the strengths and weaknesses of the program from a student's perspective. Is it a student friendly institution and a place where students receive adequate guidance? Ask about the quality of the courses and the interest the faculty take in educating the students.

A 2007 report indicates MD-PhD students are least satisfied with the thesis component of their MD-PhD training program.[1] Therefore determine as much as you can about the mentoring process, selection of a research lab, and support you will get as a student. Is the curriculum sufficiently flexible to meet the needs of students with different interests and backgrounds? If you are an ethnic or racial minority student, will you find a supportive environment at the school and community? Are courses presented in a schedule so that students can take needed courses before they take their comprehensive examinations? Do all faculty in the department or interdepartmental program actively participate in the education program? Do faculty care about educating students or do students just help increase the research productivity of their lab?

■ INTERVIEW ISSUES SPECIFIC FOR MD-PhD APPLICANTS

MD-PhD applicants may interview with faculty from the MD-PhD admissions committee and the medical school admissions committee. MD-PhD admissions committees usually include representatives from the basic science and clinical research faculty, whereas medical school admissions committees typically have a high proportion of faculty representatives from clinical departments with patient responsibilities. These two groups of faculty may have different perspectives on the type of student they would like to recruit to their school's education programs. Consequently, you need to be prepared to answer different types of questions during your interview. Members of the MD-PhD admissions committee will likely be interested in having you explain your motivation for the combined degree program. They also will be interested in all the issues we discussed previously that are in common with PhD and MD-PhD interviews, especially your passion for research.

Medical school admissions committee members will likely be interested in having you explain your motivation for medical school. They may also want you to explain how your undergraduate education and extracurricular activities contributed to the development of your character, maturity, commitment, compassion, social awareness, and empathy. You may be asked questions designed to identify your knowledge about issues related to providing healthcare delivery or about current news events (culture awareness). Interviewers will listen to your responses as well as evaluate your body language and interpersonal communication skills. Students interested in the MD-PhD may be asked additional questions to determine the applicant's potential as a physician.

Issues MD-PhD applicants should be able to demonstrate or articulate during the interview

- Demonstrate an interest in others and do not be self-absorbed.
- Instill trust and confidence.
- Be able to discuss your most significant accomplishments and why they are important.
- What are your greatest strengths and weakness? How will they affect your performance?
- Sell yourself without being arrogant.
- Be able to field questions about contemporary medical issues and advances in the field of medicine.
- Be knowledgeable about the academic program of the medical school and MD-PhD program where you interviewing.
- Articulate why you are interested in a career in medicine. How did you develop an interest in a career combining research and medicine and how is this dual career consistent with your aspirations?
- Say enough but not too much. Listen carefully to the questions and answer each part of a multipart question.
- Plan your closing comments that capture your key values, characteristics, and strengths.
- What would you contribute to the school's program? Be familiar with the institution's mission statement.

How to make MD-PhD interviews bidirectional

As we previously discussed, interviews should be bidirectional. You should ask medical school admissions committee interviewer's questions about the medical school curriculum. For example, you could ask them if they believe there is an appropriate balance between basic science and clinical courses during the first 2 years of the medical school curriculum. You also could ask how effectively medical students receive clinical training in the interviewer's clinical specialty. How is problem-based learning used? Are clinical techniques such as taking a patient's history or doing a physical examination integrated into the curriculum during the first or second year? Ask the MD-PhD committee members where their graduates go for residency training, options for research in the institution's fellowship programs, and how many of their graduates in the past 10 years are still in research careers (vs. switched to clinical practice or an alternative career). Because the transition between the end of the graduate program and third year of medical school can be challenging, ask interviewers if there is a formal process for students to review aspects of their clinical training to assist in this transition. You could ask any of the interviewers to describe the mentoring system in the clinical training and research phases of the MD-PhD program. What is currently the biggest issue among MD-PhD students?

Interviewers come into the meeting with preconceived ideas about the type of applicants they believe are appropriate for their education programs. Medical school admissions committee members want to recruit applicants who interact well with patients and other health-care providers. MD-PhD admissions committee members are primarily interested in your passion for research and do not want a student who drops the graduate school component to pursue just the MD. Both groups want to recruit mature, intelligent, personable applicants who understand the challenges of careers in research and clinical practice. You should align your answers to their questions with these primary objectives.

■ INTERVIEW ISSUES FOR INTERNATIONAL APPLICANTS

We previously discussed some of the issues admissions committees face when evaluating applications from outside the United States or Canada. If an international applicant's credentials are sufficient to get passed the initial screen, the next challenge is how to interview the applicant who may live in China, India, South Korea, Japan, Europe, South American, or somewhere else outside of the United States. Many schools ask one or more of their faculty to do a phone interview with the applicant. Optimally, if the education programs have a faculty member from the applicant's country, they may be asked to participate in the phone interview. Nearly all the issues we have discussed in this chapter on interviews are applicable to these phone interviews. Besides evaluating the applicant's understanding of science-related issues, the interviewers also evaluate the English-speaking skills of the applicant. The conundrum faced by the interviewers doing phone interviewers is to know if they are speaking to the applicant or to someone else who has good English-speaking skills and is "helping" the applicant. Some schools ask faculty at the applicant's school to participate in the interview.

■ HOW TO SAY THANK YOU

After returning home, promptly send a thank you note to each of the faculty with whom you interviewed. Two or three sentences should be used to thank them for their time. Indicate you were very impressed with the quality of their school's education program and suggest your participation in their program would help you meet your education goals. Continue to be very careful with comments placed on online chat rooms, MySpace, and Facebook. Many institutions monitor these sites, and an inappropriate entry can be very detrimental to an applicant. Some students have even had their acceptances retracted after posting an inappropriate comment.

■ SUMMARY

1. The interview is a very important component of the selection process, and you should go to as many different interviews as you receive offers.

2. The fastest way to negate the productivity resulting from 3 or more years of hard work in your undergraduate education is to not adequately prepare for a 30-minute interview.

3. You must be able to clearly articulate that you have a passion for science and, if a MD-PhD applicant, for medicine as well as science.

4. You should be able to clearly articulate the theoretical and experimental aspects of your previous research experiences.

5. You should know the research strengths of the institution where you are interviewing and be able to describe how your interests in science align with the program's strengths.

6. Write out your responses to potential questions you may be asked during an interview and practice discussing each with one of your faculty.

7. Be able to explain negative aspects of your credentials as well as discuss how you positively addressed these issues.

8. Prepare for the interview by identifying the research interests of each graduate school faculty member who may interview you.

9. During an interview, present yourself as animated, energetic, and motivated.

10. Regardless of what happened during your interview day, you must maintain a mature, positive attitude with everyone (administrative staff, faculty, and students) you encounter during your conversations.

11. Social periods can be a valuable opportunity to learn about how faculty interact with their students. You should maintain a professional, positive discussion with everyone you encounter.

12. There are illegal or inappropriate questions that interviewers may ask you. Politely decline to respond to such questions by indicating the answer to the question will not affect your performance as a medical student or graduate student.

13. MD-PhD admissions committees usually include representatives from the research faculty, whereas medical school admissions committees typically have a high proportion of clinical faculty. These two groups may have very different types of questions about your application.

14. Interviews should be bidirectional, so plan questions that you can ask concerning the school's education programs.

■ REFERENCES

1. Ahn, J., Watt, C.D., Man, L-X., Greeley, S.A, Shea, J.A. Educating future leaders of medical research: Analysis of student opinions and goals from the MD-PhD SAGE (Students' Attitudes, Goals, and Education) Survey. Academic Med. 82: 633–645. 2007.

CHAPTER

8

Appointment Processes for PhD and MD-PhD Programs

ISSUES FACULTY FOCUS ON WHEN THEY SUMMARIZE INTERVIEWS WITH EITHER PhD OR MD-PhD APPLICANTS

Faculty who participate in the PhD or MD-PhD interviews usually complete interview forms on each applicant similar to the one previously discussed in Box 6-4. Comments are made concerning the following issues:

- Can the applicant describe in detail the nature, technical aspects, and significance of their research experience?
- Did the applicant make a genuine intellectual contribution to the research project or primarily make technical contributions?
- How has the applicant demonstrated the commitment, drive, energy, organizational skills, independence, and enthusiasm for a career in research?
- Does the applicant understand the details of a career in research including effort/time commitment, writing scientific manuscripts, contributing to an education program, and writing grants to fund their research?

- Does the applicant have aspirations for a career other than laboratory investigation? If so, is this career path aligned with our program's education objectives?
- Evaluate the maturity of the applicant concerning interpersonal interactions, ability to make a career decision, and ability to meet the rigorous demands of our program.
- Is there evidence the applicant has the creativity and curiosity to be a productive investigator?
- Evaluate the applicant's communication skills including attentive listening and clearly articulating ideas.
- Assess whether the applicant will likely succeed in the academic and research components of our graduate program.
- Identify the strengths and weaknesses of the applicant.
- Rank the applicant as compared with others in the current pool of applicants or our current students.

■ ISSUES FACULTY FOCUS ON WHEN THEY SUMMARIZE INTERVIEWS WITH MD-PhD APPLICANTS

Additional comments for MD-PhD applicant interviewers may focus on issues that are specific to the MD-PhD program or to the medical school:

- Does the applicant understand the rigorous demands of the combined program?
- Does the applicant understand the competing demands of research and clinical practice in the career of a clinical investigator?
- How have the applicant's medically related extracurricular activities helped the student understand the complexities of clinical medicine?
- How has the applicant demonstrated leadership skills while working as part of a team in both research and clinical medicine?
- Does the applicant demonstrate an interest in others and have the empathy needed to be a physician?
- Is the applicant interested in becoming an MD for the right reasons?

- Will the applicant be able to multitask and manage his or her time appropriately?
- Does the applicant inspire trust and have the compassion needed to be a caring physician?
- Has the applicant clearly been able to articulate why he or she wants to be a physician as well as a clinician with an interest in research?
- Has the student demonstrated his or her ability to handle significant challenges in his or her life and does he or she have the stamina to succeed?
- Is the applicant positive and does he or she generate confidence without appearing to be arrogant? Is the applicant too driven, with the attitude of a "gunner"?
- Is the student a good match for the institution's goals and mission?

■ ADMISSIONS COMMITTEE PROCESS FOR EVALUATING APPLICANTS

When the admissions committee meets after the interviews are completed, they review the information previously summarized when interviewees were selected (see Box 6-4) plus they review faculty comments from the applicant's interviews. It is not unusual for a committee to have invited applicants to the interview who have one or more issues that could potentially exclude them from being accepted to the program. Programs may be reluctant to exclude "diamonds in the rough," because the paper credentials of applicants may not always identify students with significant potential as research investigators. Faculty have learned from experience that not all applicants are knowledgeable about the application process and may fail to highlight their strengths on the written application.

During the admissions committee meeting, faculty discuss what they perceive as the strengths and weaknesses of each applicant. Frequently, new information about the applicant that was identified during the interviews or social periods is highlighted. It is difficult to underestimate the potential positive (or negative) impact interviews can have on being accepted to a PhD or MD-PhD program. Applicants who are weak in one or more academic areas (grades, breadth of science classes, quality of school), research experience, or letters of recommendation can potentially negate such weaknesses by demonstrating their positive attributes when they meet faculty one on one. These applicants can end up highly rated by interviewers if they can articulate their passion for research, their depth of understanding of their research experiences, their humility, their basic understanding of science, and directly address the reasons for their deficiencies. On the other hand, applicants who look strong on paper can come across during their interviews as arrogant, self-centered, and, if not appropriately prepared for their interview, having a poor understanding of their research experiences and of clinical medicine. These individuals usually sink like a stone when reviewed by admissions committees because faculty do not want to deal with prima donnas in their education programs. Frequently, each admissions committee member assigns a score to each applicant based on an algorithm developed by the education program. The scores for each admissions committee member are averaged into a final score. Applicants with the highest scores are offered a position. Some programs accept more than the number of available positions because they know many applicants will receive multiple offers and only a fraction will matriculate into their program.

■ WAITING FOR THE DEGREE PROGRAM'S DECISION

After you return home after your interviews, the outcome of all your efforts in the application process is largely out of your control. Now you must wait to see if you receive an e-mail or letter of acceptance from each school where you interviewed. There may be good news or bad news, and we address these issues separately. Do not contact the school during this time unless you have significant new information to give the committee. Overanxious applicants can really turn off a committee.

First, the good news

You have been accepted at one or more schools. How do you decide which program to accept? It is time to review the issues discussed in Chapter 5. Also, review the list of issues presented in Box 5-1. Be sure to consider quality-of-life issues because you will be living for 5 to 8 years in the community where the institution is located. Rank each education program for each issue (1 = low to 5 = high) (Box 8-1 and Box 8-2). By calculating the average score for each

Box 8-1 Checklist for Selecting a PhD or MD-PhD Program

For each issue, give 1 (low) to 5 (high) points for each program you are evaluating. Include input from your academic and/or research advisors when you go through this process. More input will help you make a decision you will not regret later!

	Points
Number of research faculty (in your area of specialization)	_____
Quality of research faculty (in your area of specialization)	_____
Quality of publications	_____
Publications coauthored within department/program	_____
Investigator's extramural research grants	_____
Graduate student productivity (publications)	_____
PhD curriculum options (class requirements and options)	_____
Location	_____
Quality of life in the community	_____
Cost of living	_____
Academic reputation	_____
Student diversity	_____
Average number of years to complete degree	_____
Laboratory facilities	_____
Core facilities	_____
Stipend funding (amount and source of funding)	_____
Alignment of institution's research strengths with your interests	_____
Faculty commitment to graduate education	_____
Teaching requirements	_____
Faculty mentoring of students	_____
Computer support facilities	_____
Friendliness of students and faculty	_____
"Gut feeling"	_____
Total points:	_____

Box 8-2 Checklist for MD-PhD–Specific Issues

Medical Scientist Training Program grant in program?	_____
Reputation of medical school	_____
Medical school curriculum	_____
Quality of clinical training	_____
Alignment of institution's clinical strengths with your interests	_____
Integration of research with medical training	_____
Total:	_____
Points from Box 8-1:	_____
Total points:	_____

---

program, you will have a semiquantitative measurement of how you rank the programs. Now ask yourself how you feel about the ranking. For many students, the "emotional criteria" will be added into the ranking scores. This is a good time to discuss your ranking scores with your faculty advisor and research mentors. They can help you think through the important issues. These discussions may enable you to focus on a program that will meet your educational goals. If you have carefully identified issues that are important to you during the application process, the options before you should all be good options. You should be able to get a good education at any of the schools that have accepted you. Many PhD programs give applicants a date by which they must formally accept or reject their offer. If you decline an offer, the program may go down its list and make an offer to another applicant. It is not appropriate for you to maintain multiple acceptances from different PhD programs. If you are accepted by a program after you have accepted an offer from another school, you should quickly decline one of the offers. Ethics are a very important component in the careers of professional scientists. Do not start off the training program for your career by acting in an unprofessional manner.

MD-PhD programs realize that students with strong academic and research credentials will be accepted to more than one MD-PhD program. Because acceptances are submitted to the Association of

American Medical Colleges (AAMC) database, program directors know which schools have accepted you. Many MD-PhD programs offer you the opportunity to revisit the school so you can discuss research options with faculty and they can recruit you to their program. Your expenses for this visit are usually paid by the program. Keep in mind that this visit is not an interview, it is a recruitment trip. However, if you demonstrate inappropriate behavior during the visit, the school can withdraw your acceptance to the program. These visits can be a good opportunity to clarify issues about a program as well as meet additional faculty and students currently in the program so you can determine if these are the type of people you will want to interact with during your degree program.

The AAMC has established rules that both applicants and medical schools must follow concerning acceptance procedures for MD-PhD applicants (www.aamc.org/students/applying/policies). You should carefully read these rules. Students will have a designated time period in which to accept or reject the offer of acceptance. If you are accepted at more than one medical school, you can maintain multiple acceptances at several schools until the national deadline (for most schools it is May 15, whereas others use April 15). At this time, the applicant must select one and decline the other offers. Students may continue to remain on other school's waiting lists. If another acceptance is subsequently received, the applicant should accept one preferred school and decline the other. After May 15 (or April 15) of the matriculation year, the applicant can hold only one acceptance and must formally decline all others. Once orientation programs for MD-PhD students have begun, students can no longer accept admissions offers from other schools.

Deferring the start of your degree program

Some programs allow students to defer for 1 or more years, but the deferments are only given for significant reasons. For example, a successful Rhodes or Fulbright Fellowship or a significant personal issue such as pregnancy might be the basis of a successful request. Requests must be made in writing to the medical school or graduate school.

How to deal with the bad news

If you were not accepted to any degree program, it is time to reevaluate your application credentials. Your initial sources of information to help you reevaluate your application materials are your undergraduate college advisor and your research mentors. These people should know you and be familiar with your strengths and weaknesses. You should contact by e-mail or phone the director of the programs where you applied to ask what you could do to improve the competitiveness of your application. If the program is large and had many applicants, the program director may give you only a limited amount of useful information. However, program directors should have the application review summaries taken during the admission committee application review meeting. Talk with these individuals in person and make a written list of their suggestions (even if you disagree with their comments). An excellent resource for students in investigating why they were unsuccessful can be found in the *Pfizer Medical School Manual* (http://www.amazon.com/2007-Pfizer-Medical-School-Manual/dp/1889793213). If you decide to reapply next year, you want to review these suggestions before you rewrite your application.

Strategies to improve your credentials

To be competitive, you must address the deficiencies in your application before reapplying. Some deficiencies, such as low grade point average, are difficult to correct, whereas low Graduate Record Examination or Medical College Admissions Test scores can usually be improved with appropriate study before retaking the exam. If you are determined to reapply, consider looking for a job in science, perhaps as a research laboratory technician. Laboratory technician positions may be available at your institution or at an academic research institution near you. A 2-year laboratory experience is usually expected to be fair to your employer and to really acquire the research experience a laboratory position provides before reapplying. As discussed previously, a position with the National Institutes of Health (NIH) through the IRTA program (http://www.training. nih.gov/student/Pre-IRTA/irtamanualpostbac.asp), the Centers for Disease Control and Prevention EID Laboratory Fellowship program (http://www.cdc.gov/ncidod/eid/vol3no2/news251a.htm), or a post-baccalaureate program provide excellent opportunities to participate in academic classes and a research project to improve your application credentials. Consult the websites for the appropriate application deadlines.

In many academic institutions, you can take graduate-level classes as an employee without being accepted into a degree program. If you take one class at a time, you can focus on getting good grades in your classes. By doing well in these classes, you will demonstrate to the schools where you subsequently reapply that you have the academic ability to succeed in graduate school. Some institutions may want you to take a full academic load and will minimize high grades where only one class has been taken. Your job as a laboratory technician or post-baccalaureate student will give you additional classes, laboratory skills, and research experience that also enhances your reapplication. The head of the laboratory where you do research can provide another letter of recommendation for your application. It is important that you contribute intellectually to the projects in the laboratory and demonstrate a strong work ethic. Ideally, you also may have the opportunity to be a co-author on one or more publications, which enhances your application. Alternatively, the experience of working in a research environment may help you to decide that this career is not for you. There are many career options outside of biomedical research where you can effectively utilize your talents and have a rewarding future. It may be of value for you to make an appointment with your undergraduate school's career advising office to discuss your interests and options. Many people try a number of career options before they identify one that suits them, so take time to identify the type of activities that give you satisfaction and add value to your life. You should enjoy what you do for a living just as much as what you do in your free time.

Not being accepted to a degree program (or the program you wanted to attend) can be a frustrating and disappointing experience. Some students respond to this setback by deciding to take extended time (6 to 12 months) off to travel. Although traveling can enhance your maturity and help develop life experiences, it will *not* enhance your application to a PhD or MD-PhD program. Many directors of biomedical higher education programs view extended travel activities as a demonstration of unfocused behavior and a lack of commitment to a career as a research investigator. If you decide to take a year or more away from formal education, you should seek employment in a science or a related field (including computer science).

If you decide to reapply to PhD or MD-PhD programs, review the list of suggestions that your advisors previously gave you at the end of the previous application cycle and any feedback you have received from the PhD or MD-PhD program directors. As you write your applications, focus on changing the things that you can to enhance your application. You must be able to demonstrate that you are different from the time when you previously made an application to the advanced degree program and have addressed the issues that were negatively evaluated in your previous application. Medical schools may review your current as well as your previous application to their program. Perhaps you should apply to additional educational programs where you may have a greater chance of being accepted. Review the issues discussed previously in Chapter 5 as you reevaluate the schools where you should apply. Develop short- and long-range plans including what career options you will pursue if you are not successful with the second application.

Good luck in pursuit of your life goals!

■ SUMMARY

1. It is difficult to underestimate the potential positive (or negative) impact interviews can have on being accepted to a PhD or MD-PhD program.

2. Applicants who are weak in one or more academic areas (grades, breadth of science classes, quality of school), research experience, or letters of recommendation can negate such weaknesses by demonstrating their positive attributes when they meet faculty one on one.

3. If you have multiple acceptances, carefully rank the schools on the issues that are important to you to help you select the top school (see Box 5-1).

4. MD-PhD applicants must follow the rules established by the AAMC concerning acceptance procedures.

5. If you were not accepted to any degree program, it is time to reevaluate your application credentials by soliciting suggestions from program directors, faculty advisors, and research mentors.

6. Working for at least 2 years as a laboratory technician or in a post-baccalaureate program can help make your subsequent applications more competitive.

A

Timeline for Advanced Degree Applicants

■ FIRST YEAR OF COLLEGE

- Reflect on how you can integrate your personal interests with possible career choices. What courses do you need to prepare for these career options?

- Plan to take classes that potentially align with your interests but also try new areas, especially in science, to expand your horizons.

- Interact with your science advisor to plan a multiyear schedule for taking required course work, possible premedical courses, and other degree requirements. Pay close attention to sequential courses when you plan your academic schedule.

- Tentatively integrate any off-campus semester programs into your academic plans.

- Explore opportunities for service, volunteering, and extracurricular activities. Remember, this is your opportunity to try different career-related activities.

- Consider volunteering in a research laboratory for a few hours each week.

- Attend departmental seminars and career-related sessions that involve research as their focus.

- Identify summer employment, shadowing opportunities, or service experiences.
- Get to know your professors; remember, some of them will be writing letters of recommendation for you.

■ SUMMER

- Reflect on your first year of college. Have you developed good time-management skills? Reflect on how your first-year classes may have expanded your interests and identify potential career options that may align with your current interests.
- Conduct informational interviews or shadow professionals in areas of interest. Continue to explore new career options.

■ SECOND YEAR OF COLLEGE

- Focus on identifying a possible academic field of specialization (major). Reflect on your interests and what fits you.
- Meet regularly with your science advisor and, if medicine is a possible career choice, also meet with a pre-med advisor.
- Continue to explore career options by evaluating the components of your classes that particularly interest you.
- Attend career information sessions at regional graduate or medical schools.
- Consider an internship or shadowing experience in medicine, if that is a potential career interest.
- Attend research seminars in your science departments and information sessions held by visiting graduate school representatives.
- Volunteer! Expand your horizons by participating in appropriate service activities and extracurricular activities.
- In the fall semester, talk with your science advisor about exploring opportunities for summer research. Apply early to meet the appropriate deadlines.
- At the end of the academic year, review your academic progress. Are you meeting your goals (grade point average [GPA], science GPA, time-management skills, interacting with faculty)?

■ SUMMER

- Participate in a summer research experience. Volunteer in a local research laboratory if a paid summer research position is not available.
- If the opportunity occurs, visit graduate schools and, if interested, medical schools in the area.
- Volunteer and continue service activities.
- Attend summer school if necessary.

■ THIRD YEAR OF COLLEGE

- Explore the options for PhD and, if relevant, MD-PhD programs with your advisor.
- Acquaint yourself with the application process and the relevant timelines for advanced degree programs.
- Continue involvement in service, volunteer activities, and extracurricular activities.
- Have you developed the ability to multitask and acquired good time-management skills?
- Explore opportunities to conduct a research project on campus or to volunteer in a research laboratory.
- Attend research seminars in your science departments and graduate school information sessions.
- Apply in the fall or early in the spring for summer research opportunities in a National Institutes of Health (NIH)-funded laboratory. Attempt to find a position in a different research laboratory if your previous experience was in field research or at an undergraduate research laboratory.
- Meet regularly with your science advisor. Reflect on your career choices, your interests, and your motivators.
- Identify a list of graduate schools and, if appropriate, MD-PhD programs where you are interested in applying using the Internet, the *Medical School Admission Requirements* reference book (www.aamc.org), Peterson's Guide (http://www.petersons.com/GradChannel/code/AcdSearchResults.asp?sponsor=1), and consulting your science advisor.

■ MD-PhD APPLICANTS

- Meet with the pre-med advisor to review your advanced degree options.
- Complete the pre-med course prerequisites for the Medical College Admissions Test (MCAT) before the end of the junior year.
- Consider completing an internship or shadowing program in the medical arena.
- If available, contact your institution's pre-med committee to determine the process to obtain a committee evaluation or reference letters from faculty.
- Take a practice MCAT test in the fall. Determine your strengths and weaknesses. Initiate studying for the MCAT in January and optimally take the test in the spring.
- Contact professors and research mentors concerning potential letters of reference. Provide background material and discuss with them your career goals and graduate school choices. Provide them with a timetable of the application process and when their letters need to be submitted.
- Complete a rough first draft of the American Medical Colleges Application Service (AMCAS) essays.
- Late in the spring, consult with your advisor to finalize a list of MD-PhD programs where you will apply. Reflect on your GPA, MCAT scores, and research activities in selecting the appropriate programs.
- Late in the spring, initiate the web-based AMCAS application. Complete the applications as soon as possible.

■ SUMMER

- Immerse yourself in a summer research project in an NIH-funded laboratory.
- If a MD-PhD applicant, submit the AMCAS online application and complete the secondary applications from each MD-PhD program within 2 weeks of receiving them. Confirm that your application is complete with each institution.
- Initiate studying for the GRE. Take the exam late in the summer or early in the fall.

■ FOURTH YEAR OF COLLEGE

- If possible, continue your summer research project during the school year.

- Continue going to research seminars in your science departments and graduate school information sessions.

- Explore graduate fellowship and scholarship programs with your advisor. If you fit the fellowship profile, complete an application and prepare a research proposal with the help of your research mentor.

- Based on your GPA, GRE scores, and research background, interact with your advisor and research mentors to finalize your list of graduate schools of interest.

- Complete the application to each institution.

- Request letters of recommendation from professors and research mentors for each application.

- Contact each graduate school to ensure that your application is complete and all the letters of recommendation have arrived.

- When invited for an interview, respond immediately and arrange an interview time. Carefully prepare for each interview.

- Send thank you notes after each interview.

- Patiently wait until each school contacts you.

- If you are fortunate to receive multiple acceptances, you might visit the programs again to help in making your decisions.

- For MD-PhD applicants, you must select one program by May 15. Graduate schools have individual acceptance deadlines. Send letters of regret to each of the other schools.

- If you are on an alternate list, make sure the school will be able to contact you or your parents if a position becomes available.

- If you were not successful at acquiring a position, meet first with your advisor to determine your next steps and to reflect on your qualifications. You can also contact the admissions departments of the schools for additional feedback.

- Consider a plan of action if not accepted the first time. How can you improve your application and profile?

- Enjoy your senior year! Celebrate a successful undergraduate experience.

APPENDIX

B

Book and Website References

■ REFERENCE BOOKS

Writing applications for advanced degree programs

Get into Graduate School: A Strategic Approach (Get into Graduate School) (Paperback). 2003. Publisher: Kaplan. By Kaplan.

This book includes advice on writing persuasive personal statements, obtaining the recommendations, how to prepare your application, information on financial aid, borrowing, and managing expenses. It provides specific information for minority students, older students, people with disabilities, and international students. There are resources to help graduate school applicants find graduate programs, and the book discusses what to include and not include in your application.

Graduate Admissions Essays: Write Your Way into the Graduate School of Your Choice (Graduate Admissions Essays) (Paperback). 2008. Publisher: Ten Speed Press. By Donald Asher.

This book helps students organize their application essays. The book provides instructions and exercises on how to write admissions essays. Examples of bad essays are provided as writing pitfalls that you should avoid.

Getting What You Came For: The Smart Student's Guide to Earning a Master's or a Ph.D. (Paperback). 1997. Publisher: Farrar Straus Giroux. By Robert Peters.

This has advice for both applicants and students in graduate programs. Information is provided for applicants concerning how to apply to graduate school and how to obtain financial aid. Suggestions are provided to graduate students on how to excel on qualifying examinations and how to manage academic politics (including hostile professors) and how to write and defend your thesis. The book also provides information on how to land a job when you graduate.

How to Write a Winning Personal Statement for Graduate and Professional School (Paperback). 1997. Publisher: Peterson's. By Richard J. Stelzer.

For students who seek to gain admission to law, business, or medical school. This book gives advice about how to write an application essay, including how to start, what to include, and what you should never say. The book provides examples of personal statements, and what admissions directors are looking for.

Graduate School: Winning Strategies For Getting In With Or Without Excellent Grades (Paperback). 2004. Publisher: Proto Press Publications. By Dave G. Mumby.

This is an easy-to-understand book that takes you though steps needed to convey your unique qualifications in a graduate school application. The author provides insights concerning what admissions committees are looking for when assessing a student's application. The book discusses how to obtain quality letters of recommendation, the importance of cultivating preapplication contact with prospective supervisors, and writing a personal statement.

Preparing for the GRE

Verbal Workout for the GRE, 2nd Edition *(Graduate Test Prep)* (Paperback). 2007. Publisher: Princeton Review. By Princeton Review.

This book helps you prepare for the verbal portion of the GRE test. It includes hundreds of practice exercises to improve your skills, solve analogies without knowing all the words in the problem, contextual clues to answer sentence-completion problems, manage your test time on the reading comprehension section, and provides a list of the 300 vocabulary words that most frequently appear on the exam.

Cracking the GRE with CD-ROM, 2006 (Graduate Test Prep) (Paperback). 2005. Publisher: Princeton Review. By Princeton Review.

This guide teaches techniques to help students prepare for the computerized format of the Graduate Record Examination. It prepares students to solve analogies and become prepared for the problems in the Verbal and Quantitative sections. Book buyers get a free pass to go online and take full-length practice tests.

How to Prepare for the GRE with CD-ROM (Barron's How to Prepare for the GRE Graduate Record Examination), 16th Edition (Paperback). 2006. Publisher: Barron's Educational Series. By Sharon Weiner Green and Ira Wolf.

This book presents six full-length tests similar to the actual GRE in length and difficulty with questions answered and explained. The manual reviews GRE test topics that include antonym questions, analogy questions, sentence completion, reading comprehension, vocabulary, analytical writing, quantitative comparison, data interpretation, and math. An overview of the GRE is presented with advice for study preparation and test taking. A 3,500-word master vocabulary list with definitions is supplemented with a GRE high-frequency word list. The CD-ROM enclosed with the book offers two practice GRE tests that are similar in structure to the actual GRE test, including automatic scoring.

GRE Exam 2008, Premier Program (Kaplan GRE Exam) (Book and CD-ROM) (Paperback). 2007. Publisher: Kaplan. By Kaplan.

This guide offers preparation tools such as a personalized online study plan that delivers new practice questions every month, a CD-ROM with additional questions and tests, and study materials that can be downloaded to personal data assistants and cell phones. The book provides score-raising strategies and five full-length practice tests (one in the book, one online, and three on CD-ROM). More than 300 additional questions are included with detailed answer explanations.

Kaplan GRE Exam Verbal Workbook (Kaplan GRE Verbal Workbook), Fourth Edition (Paperback). 2006. Publisher: Kaplan. By Kaplan.

This book offers test takers vocabulary-building methods to help prepare for the GRE Verbal section. This book includes hundreds of practice questions with detailed answer explanations, score-raising strategies, a mini-dictionary of more than 1,000 words that test takers should know, and the new question types on the most recent version of the GRE. In addition, there is a chapter dedicated to the Analytical Writing Section.

Kaplan GRE Exam, 2007 Edition: Comprehensive Program (Kaplan GRE Exam) (Paperback). 2006. Publisher: Kaplan. By Bruce Simmons.

This book includes review for all content on the GRE and incorporates Kaplan's score-raising strategies. Features include two full-length practice tests (one in the book, one online) with detailed answers and explanations; 300 additional practice questions; and Kaplan's strategies for achieving a higher score, including strategies for increasing Reading Comprehension, GRE Exam Writing, and Problem Solving questions.

Word Smart for the GRE (Smart Guides) (Paperback). 2007. Publisher: Princeton Review. By Anne Curtis.

This book is designed to help you build your vocabulary. It gives strategies for learning new words and for improving your scores on the GRE Verbal exam. The book includes words most frequently tested on the exam as well as quizzes and a final exam. It provides a list of roots with examples of words that contain each of them, and tips for incorporating roots into your vocabulary regimen.

The GRE Test for Dummies (Lifestyles paperback). 2006. Publisher: For Dummies. By Suzee Vilk, Michelle Gilman, and Veronica Saydak.

This paperback includes strategies to prepare you for the computerized version of the GRE test. The book includes basic information about the test, strategic advice as to how to approach the exam, as well as sample questions and practice exams.

GRE: Practicing to Take the General Test (Paperback). 2002. Publisher: Educational Testing Service. By Educational Testing Service.

This book contains questions and topics from actual tests administered by Educational Testing Service. It contains verbal and quantitative questions from seven GRE general tests, sample analytical writing topics from topics for the analytical writing exam, and test-taking strategies. It includes information about the structure of the test, answering procedures, answers and explanations for verbal and quantitative questions, sample writing responses with scores, scoring information, and a math review.

Preparing for TOEFL exams

TOEFL iBT: The Official ETS Study Guide (McGraw-Hill's TOEFL iBT) (Paperback). 2005. Publisher: McGraw Hill. By Educational Testing Service.

Information about the TOEFL® iBT from the company that generates the exam. This book includes hundreds of authentic TOEFL iBT questions. It has a companion audio CD with authentic TOEFL listening passages and sample responses to speaking questions plus descriptions and answers for the multiple-choice Listening and Reading questions, with valuable tips for answering them. Information is provided about the Speaking and Writing sections, with scoring information, real student responses, and raters' comments. Included is the *Writer's Handbook for English Language Learners* to help you learn how to write essays in English.

Longman Preparation Course for the TOEFL® Test: Next Generation (iBT) with CD-ROM and Answer Key (Longman Preparation Course for the TOEFL) (Paperback). 2005. Publisher: Prentice Hall. By Deborah Phillips.

This book/CD-ROM package provides the tools for the new TOEFL® integrated skills test. It presents both a language skills course and practice for all sections of the test. It has diagnostic pre- and post-tests, including eight mini-tests and two complete practice tests.

TOEFL iBT with CD-ROM (Kaplan TOEFL Cbt) (Paperback). 2007. Publisher: Kaplan Publishing. By Kaplan.

This guide explains the format of the exam in detail and has a CD-ROM with sample tests and answer keys. It has an integrated review of the integrated English language skills and strategies to help prepare for the exam.

Delta's Key to the Next Generation TOEFL Test: Six Practice Tests for the iBT (Paperback). 2006. Publisher: Delta Systems Company. By Nancy Gallagher.

This book contains practice materials for students preparing to take the Test of English as a Foreign Language. The tests contain questions that are similar in form and content to the questions on the TOEFL iBT and TOEFL ITP (institutional TOEFL). The book provides strategies for practicing essential skills and for taking the exam. In addition to the tests, the book has six full-length practice tests, over 600 questions in the format of the next generation TOEFL, and answer keys for the questions. An optional audio CD is available that offers 6 hours of listening on six CDs to help prepare for the listening, speaking, and writing sections of the exam.

Building Skills for the TOEFL iBT (Advanced Student Book with Audio CDs) (Paperback). 2005. Publisher: Pearson ESL. By Linda Robinson Kellag.

The book contains 10 thematic units to develop skills in English, critical thinking, and communication. It includes authentic TOEFL® iBT practice sets created by ETS especially for this series and strategies for taking the test. A teacher's manual provides evaluation tools to track students' progress and actual student responses (speaking and writing) at all score levels.

Cracking the TOEFL iBT with Audio CD, 2008 Edition (College Test Prep) (Paperback). 2007. Publisher: Princeton Review. By Princeton Review.

This book contains a full-length practice test and many additional practice questions with an explanation of the answers. It includes strategies and test-taking techniques to prepare for the exam, including how to focus your reading and listening to identify key parts of passages, lectures, and conversations. One section is designed to increase your skills to write essays by responding to the specific question asked and by helping you organize your ideas. An audio CD is also included.

Preparing for the MCAT

MCAT Practice Tests, Fourth Edition (Paperback). 2004. Publisher: Kaplan Publishing. By Kaplan.

This book contains one full-length practice test and two practice tests for each of the four sections of the MCAT (Biological Sciences,

Physical Sciences, Verbal Reasoning, and Writing) plus answers and explanations for the questions. It also presents Kaplan's strategies for taking the exam.

The Gold Standard MCAT (Paperback). 2008. Publisher: Ruveneco Inc. By Brett Ferdinand.

Complete and updated review for the new Medical College Admissions Test (changed in 2008). It includes a review of the MCAT, with three full-length practice exams with answers and explanations. Free online access is provided to full-length exams on the publisher's website. This manual covers different aspects of medical school admissions process, including how to prepare for medical school interviews, how to write personal statements, and how to obtain recommendations. This author also published two companion DVDs: The Gold Standard Video MCAT Science Review on four DVDs on physics and on biology (2007).

Arco Gold MCAT Sample Exams (Academic Test Preparation Series) (Paperback). 2004. Publisher: Arco. By Stefan Bosworth, Marion Brisk, and Ronald P. Drucker.

This book contains three full-length sample MCAT exams with complete answer explanations.

MCAT Verbal Reasoning Review (Arco MCAT Verbal Reasoning Review) (Paperback). 2000. Publisher: Arco. By Stefan Bosworth and Rosie M. Soy.

This book contains strategies for answering the verbal reasoning questions on the MCAT. It contains four full-length practice exams and detailed answer explanations for each question. It also contains strategies to help you improve your responses on the MCAT writing sample.

Peterson's MCAT Success 2005 (Peterson's MCAT Success) (Paperback). 2004. Publisher: Peterson's. By Stefan Bosworth, Marion A. Brisk, Ronald P. Drucker, Denise Garland, Edgar M. Schnebal, and Rosie Soy.

This book offers reviews for the Verbal Reasoning, Physical Sciences, Biological Sciences, and the Writing Sample components of the MCAT. This book offers a review of the science to those students who may have forgotten the information that they learned in their early college years.

Practice MCATs (Graduate Test Prep), Second Edition (Paperback). 2007. Publisher: Random House Information Group. By Princeton Review and Theodore Silver.

This book contains hundreds of practice questions and provides access to online practice tests that are similar to what you will encounter on the MCAT examination. The concepts addressed in the questions are reviewed and explained. There is a section on the MCAT essays as well as a review of the mathematics, physical and biological sciences, and verbal skills needed for the exam.

Developing interviewing skills

Fearless Interviewing: How to Win the Job by Communicating with Confidence (Paperback). 2002. Publisher: McGraw-Hill Companies. By Marky Stein.

This book provides a strategic approach to different aspects of the interviewing process. Contains useful advice on all aspects including how to impress interviewers at the outset of the interview, how to address interview questions, and how to practice for an interview.

Crisp: Preparing for the Behavior-Based Interview: How to Get the Job You Want (Crisp 50-Minute Book) (Paperback). 2001. Publisher: Crisp Learning. By Terry L. Fitzwater.

This book presents suggestions on how to prepare for an interview in which you will be asked to describe your experience with tasks valued by potential employers.

The Interview Rehearsal Book (Paperback). 1999. Publisher: Berkley Trade. By Deb Gottesman and Buz Mauro.

This book discusses how to develop the self-confidence, verbal communication, and body language that will help you with interviews. It also suggests how to deal with interview stress, and presents exercises for effective verbal and physical communications.

■ WEBSITE REFERENCES

Information concerning careers in biomedical research and medicine

1. https://services.aamc.org/Publications/showfile.cfm?file=version87.pdf&prd_id=198&prv_id=239&pdf_id=87
   Learning objectives for medical student education

2. https://services.aamc.org/Publications/index.cfm?fuseaction=
Product.displayForm&prd_id=121&prv_id=137&cfid=1&
cftoken=67ACC861-AA78-4802-8C07E409CBE8C058
Medical education costs and student debt

3. http://www.sciencemag.org/cgi/reprint/317/5840/966.pdf
Careers in translational research

4. http://www.sciencemag.org/cgi/reprint/298/5591/40.pdf
Article describes the intense competition for federal research
grant funding in 2008

5. http://www.sciencemag.org/cgi/reprint/304/5678/1830.pdf
Analysis of research career options and benefits

6. http://www.aspiringdocs.org/site/c.luIUL9MUJtE/b.2011109/
k.4D51/Thinking_About_Medicine.htm
Issues to think about concerning careers in medicine

7. http://www.aspiringdocs.org/site/c.luIUL9MUJtE/b.2011135/
k.BF6C/Preparing_for_Medical_School.htm
Preparing for medical school

8. http://services.aamc.org/currdir/section2/schematicManager
Remote.cfm
The AAMC website that provides curriculum information on
courses offered in medical school years 1 through 4 at each U.S.
medical school

9. http://www.usmle.org/Examinations/step1/step1.html
Information on Step 1 of the U.S. Medical Licensing Examination. This examination is given to medical students at the end
of their second year of medical school. It evaluates if students
understand and can apply important concepts of the sciences
basic to the practice of medicine.

10. http://www.physicianscientists.org/
The American Physician Scientists Association (APSA) is a national organization dedicated to addressing the needs of future
physician scientists with respect to their training and career
development.

Summer undergraduate research experience (SURE) opportunities

1. http://www.training.nih.gov/student/index.asp

   Summer programs at the National Institutes of Health (NIH) provide an opportunity to spend a summer working with some of the leading scientists in the world in an environment devoted exclusively to biomedical research. The NIH consists of the Hatfield Clinical Research Center and more than 1,200 laboratories located on the main campus in Bethesda, Maryland, as well as in Baltimore and Frederick, Maryland; Research Triangle Park, North Carolina; Phoenix, Arizona; Hamilton, Montana; and Detroit, Michigan.

2. http://www.dep.anl.gov/p_undergrad/summer.htm

   This site identifies basic science research programs at the Argonne National Laboratory in physical and life sciences, mathematics, computer science, and engineering. Also available are applied research programs relating to energy, conservation, environmental impact and technology, nanomaterials, and advanced nuclear energy systems.

3. http://www.whoi.edu/page.do?pid=7802

   Summer research programs for undergraduates, including scholarships for minorities, at Woods Hole Oceanographic Institution

4. http://www.aamc.org/members/great/summerlinks.htm

   AAMC Graduate Research, Education, and Training (GREAT) summer undergraduate research programs

5. http://www.igert.org/summer.asp

   A partial list of summer undergraduate research programs funded by the National Science Foundation (NSF)

6. http://services.aamc.org/summerprograms

   A database resource to help undergraduates locate enrichment programs on medical school campuses

7. http://www.nsf.gov/crssprgm/reu/index.jsp

   National Science Foundation Research Experiences for Undergraduates (REU). NSF funds research opportunities for undergraduate students through its REU sites program. An REU site consists of a group of 10 or so undergraduates who work in the research programs of the host institution. Each student is associated with a specific research project, where he or she works closely with the faculty and other researchers. Students are granted stipends and, in many cases, assistance with housing and travel. Undergraduate students supported with NSF funds must be citizens or permanent residents of the United States or its possessions. An REU site may be at either a U.S. or foreign location.

8. http://www.training.nih.gov/student/

   National Institutes of Health summer programs offered each summer for minority students to explore career options

9. http://sciencecareers.sciencemag.org/career_development/previous_issues/articles/2007_07_06/caredit_a0700095

   Article describes the value of research experience to gain entrance to top-tier graduate schools

10. http://www.columbia.edu/cu/biology/ug/intern.html

    Extensive database of summer internships in biology, biomedical research, ecology, environmental studies, marine biology, and for students from disadvantaged backgrounds or minorities

11. http://cns.utexas.edu/hpo/summer.asp

    Extensive database of summer internships in Texas and other states

Data from MD-PhD applicants

http://www.aamc.org/research/dbr/mdphd/Garrison.pdf

AAMC data on students applying to MD-PhD programs

Identifying PhD and MD-PhD degree programs

1. http://services.aamc.org/currdir/section3/degree2.cfm

   AAMC list of different types of combined degree programs

2. http://services.aamc.org/currdir/section3/start.cfm

   AAMC directory of different combined degree programs (sort program by school or school by program). The directory includes MD/JD, MD/MBA, MD/MPH, and MD-PhD programs.

3. http://www.petersons.com/GradChannel/code/AcdSearch Results.asp?sponsor=1

   Peterson's website identifies academic institutions that offer PhD degrees in biomedical science

4. http://www.aamc.org/research/dbr/mdphd/programs.htm
   Website that provides a list of all MD-PhD programs in the United States and Canada

5. http://www.nigms.nih.gov/Training/InstPredoc/PredocOverview-MSTP.htm
   National Institutes of Health Medical Scientist Training Program (MSTP) grants that provide training grants for highly competitive MD-PhD programs

6. http://www.lcme.org/standard.htm
   Liaison Committee on Medical Education (LCME) accreditation review information for medical schools

Information for comparatively evaluating PhD and MD-PhD education programs

1. http://portal.isiknowledge.com/portal.cgi/jcr?Init=Yes&SID=A115J21NPkJKeeOmpMP
   Thompson's Journal Citation that ranks different science journals

2. http://crisp.cit.nih.gov/crisp/crisp_query.generate_screen
   NIH website that will enable you to search for the research funding history of individual investigators

3. http://grants.nih.gov/grants/award/trends/FindOrg.cfm
   Federal grant award data for individual research institutions

4. http://www.aamc.org/data/gq/questionnaires/start.htm
   AAMC website that can be used by MD-PhD applicants to obtain information about specific medical school programs

5. http://www.aamc.org/data/facts/start.htm
   AAMC website with useful information concerning the GPA scores, MCAT scores, ethnicity, sex, and state of residency of recent medical school applicants

6. http://www.pubmedcentral.nih.gov/
   National Library of Medicine's PubMed data base where faculty research publications can be identified

7. http://grants.nih.gov/grants/award/trends/FindOrg.cfm
   National Institutes of Health Office of Extramural Research. The NIH tracks its funding of critical medical research and other

support at universities, hospitals, small businesses and other organizations, and annually compiles this information and makes it available to the public.

8. http://services.aamc.org/tsfreports/select.cfm?year_of_study= 2008

Tuition and student fees for first-year medical school students in 2007–2008

9. http://www.aamc.org/data/msq/allschoolsreports/msq2006.pdf

Matriculating student questionnaire. The web-based questionnaire asks first-year medical students to share their thoughts on a variety of topics, such as pre-med experience, opinions about the medical school application process, and reasons for choosing medical school.

Information for preparing applications to PhD or MD-PhD programs

1. http://www.aamc.org/students/applying/policies/

AAMC recommendations for medical school admissions officers and medical school applicants. AAMC rules that both applicants and medical schools must follow concerning acceptance procedures.

2. http://www.petersons.com/GradChannel/code/AcdSearch Results.asp?sponsor=1

Peterson's guide to graduate school programs. This is a graduate school database of accredited graduate programs. The graduate school lists can be searched by masters degree or doctorate degree, by graduate program, or by institution name. Once you've found a set of graduate schools that interest you, you'll have the opportunity to read more about each one and then contact admissions officials directly for more information.

3. http://www.nigms.nih.gov/Training/InstPredoc/PredocOverview-MSTP.htm

Medical Scientist Training Program (MSTP) institutions

4. https://services.aamc.org/Publications/index.cfm?fuseaction= Product.displayForm&prd_id=226&prv_id=276&cfid=1 &cftoken=68A2154C-7143-48EE-898B86F1AF916319

AAMC's current *Medical School Admission Requirements* (MSAR) can be purchased on the AAMC website.

5. http://www.aamc.org/students/minorities/resources/medmar.htm

   AAMC Medical Minority Applicant Registry. Med-MAR was created to enhance admission opportunities for groups currently underrepresented in medicine. The program's registry distributes web-based basic biographical information about the examinee and the examinee's MCAT scores to minority affairs and admission offices of AAMC member schools and certain health-related agencies interested in increasing opportunities for students who self-identify as underrepresented or economically disadvantaged individuals.

6. http://www.aamc.org/data/msq/start.htm

   AAMC website containing helpful information for writing medical school applications. The site contains responses first-year medical students gave to a variety of questions concerning their pre-med experience, the medical school application process, reasons for selecting a medical school, and future career goals.

7. https://services.aamc.org/Publications/index.cfm?fuseaction=Catalog.displayProductList&queryType=TC&chr_id=22&cfid=1&cftoken=67ACC861-AA78-4802-8C07E409CBE8C058

   MCAT computer-based practice tests

8. http://www.aamc.org/students/mcat/start.htm

   Medical College Admissions Test website

9. http://www.aamc.org/mcat

   Information on preparing for the exam, exam content, fees, dates and deadlines, and registering to take the MCAT

10. http://www.aamc.org/data/start.htm

    Student surveys and data from AAMC

11. http://www.aamc.org/students/amcas/start.htm

    American Medical College Application Service (AMCAS). This centralized application service provides the primary application that is used for all MD-PhD programs.

12. http://www.aamc.org/data/gq/start.htm

    The Medical School Graduation Questionnaire (GQ) is a national questionnaire administered by the Association of

American Medical Colleges (AAMC). It has been administered annually since 1978 to U.S. graduating medical students.

13. http://www.aamc.org/members/great/mdphd/start.htm

    Recent information concerning changes in the MD-PhD application process

14. http://www.aamc.org/members/great/

    GREAT Group (Group on Graduate Research, Education, and Training). The GREAT Group provides professional development to and fosters the exchange of information and ideas among the faculty and administrative leaders of biomedical PhD, MD-PhD, and postdoctoral programs. The group functions as a national forum to help these programs achieve their goal of educating successful biomedical researchers.

15. http://www.ets.org/gre

    Graduate Record Examination website with information about how to apply for the GRE, the GRE content, and how to prepare for the exam

16. http://www.aamc.org/data/facts/start.htm

    AAMC FACTS—Applicants, Matriculants, and Graduates. The 27 FACTS tables provide the most comprehensive and objective information regarding medical school applicants, matriculants, enrollment, and graduates available to the public free of charge.

17. http://www.aamc.org/students/amcas/amcas2007 instructions.pdf

    AAMC 2007 Student Application help web page. Detailed instruction to completing the American Medical College Application Service forms.

18. http://www.aamc.org/members/great/mdphd/start.htm

    Web page of the MD-PhD section of the GREAT Group. The mission of the MD-PhD section is to advance the education, training, and career development of physician-scientists, with an emphasis on training in the MD-PhD programs of LCME accredited medical schools.

19. http://www.amazon.com/2007-Pfizer-Medical-School-Manual/dp/1889793213

    The *Pfizer Medical School Manual*. For anyone considering medical school, this concise manual covers the entry process from initiation of applications and taking exams to interviewing and selection. Updated annually, it offers suggestions for success and statistical profiles of every medical school in the United States as well as other resources. An excellent resource for students investigating why their application to a PhD or MD-PhD degree program was unsuccessful.

20. http://www.ets.org/bin/getprogram.cgi?test=toefl&redirect=format

    Test of English as a Foreign Language (TOEFL) website. Information on preparing for the exam, exam content, fees, dates and deadlines, and registering to take the TOEFL exam.

21. http://www.pubmedcentral.nih.gov/

    PubMed Central (PMC) is the U.S. National Institutes of Health (NIH) free digital archive of biomedical and life sciences journal literature.

22. http://www.aamc.org/students/applying/policies

    AAMC recommendations for medical school admission officers and medical school applicants. The AAMC offers recommendations to ensure that applicants are afforded timely notification of the outcome of their medical school applications and have timely access to available first-year positions and to ensure that schools have no unfilled positions in their entering classes. These recommendations are distributed for the information of prospective medical students, their advisors, and medical school personnel.

23. http://www.nature.com/nature/journal/v447/n7146/full/447791a.html

    Article about how having a good mentor early in your career can mean the difference between success and failure in any field

Preparing for interviews

1. http://www.cs.wvu.edu/jobs/iviewsituation.html
   Preparing for situational interviews

2. http://www.tc.umn.edu/~bmes/assets/Sample_Behavioral_Int_Questions.doc
   Sample behavioral interview questions

3. http://www.interviewfeedback.com
   Resources and information on interview skills and interview tips

Graduate fellowship opportunities

1. http://www.nsf.gov/funding/

   The National Science Foundation promotes and advances scientific progress in the United States by competitively awarding grants and cooperative agreements for research and education in the sciences, mathematics, and engineering.

2. http://blogs.asee.org/fellowships/general-student-fellowships-and-scholarships

   National Science Foundation Graduate Research fellowship program salaries

3. http://www.grants.nih.gov/grants/guide/notice-files/NOT-OD-07-057.html

   National Institutes of Health Ruth L. Kirschstein National Research Service Award (NRSA) stipends effective for fiscal year 2007

4. https://www.grad.uiuc.edu/fellowship/category/11

   The website of the Graduate College at the University of Illinois at Urbana-Champaign has an extensive list of fellowships for science/engineering study, women, and underrepresented minorities.

Fellowship opportunities for minority students

1. http://www.aauw.org/education/fga/fellowships_grants/index.cfm

   Fellowships, grants, and awards for U.S. and international women scholars

2. http://www.uncf.org/merck/programs/grad.htm

   United Negro College Fund Merck Undergraduate Science Research Scholarship Awards

3. http://www.nigms.nih.gov/Research/Application/GrantAppRev.htm

   National Institute of General Medical Sciences Minority Access to Research Careers program. Fellowship program for underrepresented minority students.

4. http://www.aamc.org/students/minorities/start.htm

   Information for racial or ethnic minority students that are underrepresented in research careers

5. http://www.aamc.org/diversity

   AAMC diversity website providing information on initiatives for minority students

6. http://www.aamc.org/students/minorities/start.htm

   AAMC minorities in medicine website. This site provides information related to minority medical student preparation, the medical education pipeline, and financial aid opportunities.

7. https://ugsp.nih.gov/

   National Institutes of Health Undergraduate Program for individuals from disadvantaged backgrounds

8. http://www.gmsp.org

   Gates Millennium Scholars program for individuals from disadvantaged backgrounds funded by the Bill and Melinda Gates Foundation. The goal of GMS is to promote academic excellence and to provide an opportunity for outstanding minority students with significant financial need to reach their highest potential.

9. http://sciencecareers.sciencemag.org/funding

   GrantsNet is a resource to find funds for undergraduates in specific science disciplines and for different types of programs.

10. http://services.aamc.org/summerprograms

    AAMC Enrichment Programs on the web. A database resource to help undergraduates locate enrichment programs on medical school campuses.

11. http://www.hhmi.org/grants/office/undergrad

    Howard Hughes Medical Institute Undergraduate Science Education Programs. HHMI seeks to strengthen the quality of bioscience education for all undergraduate students, including non-science majors, and better prepare students for careers in biomedical research, medicine, and science teaching.

12. http://www.imdiversity.com/Villages/Channels/grad_school/gs_fellowcontents.asp#supportmed

    The Guide to Graduate and Professional School Fellowships. A list of websites identifying scholarships for minority students.

13. http://www.gradschools.com/diversity/fellowships.html

    A list of websites identifying fellowships for minority students, including portable fellowships. Portable fellowships are externally funded and allow recipients to use the funding at the institutions of their choice.

Post-baccalaureate program information

1. http://www.naahp.org/resources_postbac.htm

   The National Association of Health Professions provides a useful description of different types of post-baccalaureate programs.

2. http://www.services.aamc.org/postbac/

   The AAMC provides a searchable database for schools that have post-baccalaureate pre-medical programs. The database contains information on each program's length, size, purpose, structure, cost, admission requirements, and other characteristics.

3. http://hpap.syr.edu/pblist.htm

   This site has lists of post-baccalaureate programs directed at students underrepresented in the health professions, students who need to improve their credentials as well as those who have little or no sciences, and schools with masters programs.

4. http://www.cse.emory.edu/sciencenet/undergrad/post-bac.html

   This site provides lists of post-baccalaureate programs with a pre-med focus or health sciences/research focus.

Career and research opportunities for college graduates

1. http://www.training.nih.gov

   Description of federal research positions where students can participate in a research lab for 1 to 2 years after completion of their undergraduate degree.

2. http://www.aphl.org/fellowships/eid/Pages/default.aspx

   Association of Public Health Laboratories. The Emerging Infectious Diseases (EID) Laboratory Fellowship Program, sponsored by APHL and CDC, trains and prepares scientists for careers in public health laboratories and supports public health initiatives related to infectious disease research. The EID Advanced Laboratory Training Fellowship is a 1-year program designed for bachelor's or master's level scientists, with emphasis on the practical application of technologies, methodologies, and practices related to emerging infectious diseases.

3. http://www.training.nih.gov/student/Pre-IRTA/previewpostbac.asp
   National Institutes of Health Postbaccalaureate Intramural Research Training Award (IRTA) Program. The post-baccalaureate IRTA program and the National Cancer Institute's CRTA program provide opportunities for recent college graduates to spend a year engaged in biomedical research at the NIH. U.S. citizens or permanent residents who have received a bachelor's degree from an accredited U.S. college or university and who have held the degree for less than 2 years are eligible to apply. Trainees work side-by-side with some of the leading scientists in the world in an environment devoted exclusively to biomedical research. Fellowships are available in the more than 1,250 intramural laboratories of the NIH, which are located on the main NIH campus in Bethesda, Maryland, as well as in Baltimore and Frederick, Maryland; Research Triangle Park, North Carolina; Phoenix, Arizona; Hamilton, Montana; and Detroit, Michigan.

4. http://see.orau.org/
   Education programs at the Oak Ridge Institute for Science and Education

5. http://mayoweb.mayo.edu/mshs/
   The Mayo School of Health Sciences provides descriptions of many career options in the health sciences.

Science salaries and research funding

1. http://www.sciencemag.org/cgi/reprint/304/5678/1830.pdf
   Science magazine 2004 survey of salaries in science careers

2. http://www.sciencemag.org/cgi/reprint/298/5591/40.pdf
   Science magazine article on National Institutes of Health grant funding to young research investigators

GLOSSARY

American Medical Colleges Application Service (AMCAS) is a non-profit, centralized application processing service for applicants to the first-year entering classes at participating U.S. medical schools (http://www.aamc.org/students/amcas/start.htm).

Association of American Medical Colleges (AAMC) is a not-for-profit organization that represents the medical community on issues of medical education, research, and health care (http://www.aamc.org/about/).

Core curriculum (science) is a series of biomedical courses identified as essential for the development of an appropriate science background for students in biomedical science.

Doctor of Medicine (MD) is a degree that typically takes 4 years (in a U.S. medical school) and provides students training in the science and art of medicine for the diagnosis and treatment of disease and injury.

Grade point average (GPA) is calculated by dividing the total amount of grade points earned by the total amount of credit hours attempted.

Graduate assistantships are part-time employment where the student is paid a salary (stipend), given a work schedule and assigned responsibilities, and receives graduate tuition remission. Responsibilities frequently include either teaching or participating in assigned research projects.

Graduate fellowships support an individual during their PhD degree program. They typically provide a stipend, tuition, and healthcare benefits but do not have a formal service commitment. A variety of public and private sources provides funding of fellowships for graduate students.

Graduate Record Examination (GRE) is a test that measures verbal reasoning, quantitative reasoning, and critical thinking and analytical writing skills.

The Graduate Record Examination Subject Tests gauge undergraduate achievement in eight specific fields of study (biochemistry, cell and molecular biology, biology, chemistry, computer science, literature in English, mathematics, physics, psychology).

Health Professions (Pre-med) committee prepares letters of recommendation for undergraduate students who are applying for admission to medical, osteopathic, dental, podiatric, or optometry school. Typically, the pre-med committee prepares a letter of recommendation that summarizes a number of recommendation letters provided by a school's faculty who know the applicant.

Juris Doctor (JD) or Doctor of Jurisprudence degree is the professional training necessary to become qualified as a lawyer.

Laboratory rotations involve graduate students working in a research laboratory for a period of time that could be several weeks to several months in length. During this experience, students participate in a research project in the laboratory. This is an opportunity for a new graduate student to become exposed to the process of research, learn laboratory techniques, and to get to know the culture (personality) of the laboratory. Typically, graduate students participate in two to four of these rotations before they select a mentor for their PhD thesis project.

Liaison Committee on Medical Education (LCME) is the accrediting authority for medical education programs leading to the MD degree in U.S. and Canadian medical schools (http://www.lcme.org/).

Master of Business Administration (MBA) is an education program that provides students training in a variety of subjects such as economics, marketing, accounting, finance, operations management, international business, information technology, management, and ethics.

Master of Clinical Epidemiology (MCE) is a graduate degree that prepares students to answer clinical questions relevant to the daily practice of medicine and other health sciences and to improve patient care. Typically, programs of study include biostatistics, epidemiology, research methods in controlled clinical trials, clinical epidemiology, chronic disease epidemiology, infectious disease, advanced statistical methods for clinical research, health risk management, and injury epidemiology and prevention.

Master of Divinity (MDiv) is a professional degree in divinity that is a prerequisite for licensing to professional ministry. Coursework typically includes studies in the New Testament, theology, philosophy, church history, and the Hebrew Bible (Old Testament).

Master of Health Administration (MHA) is a graduate degree that trains students for careers involving the management of hospitals and other health services organizations. Programs of study typically involve course work in healthcare economics and policy, management of healthcare organizations, healthcare marketing and communications, human resources, information systems, operations assessment, governance, statistical analysis, and financial analysis.

Master of Health Infomatics (MHI) is a graduate degree program to train professionals as liaisons to bring a background of medicine and knowledge of information technology to solve healthcare problems. Programs typically include courses in health informatics, health administration, statistics, and research design.

Master of Public Health (MPH) is a graduate degree that trains students in applied public health practice. Programs of study typically include health services administration and management, biostatistics, epidemiology, behavioral science, and occupational and environmental health sciences.

Master of Science in Public Health (MSPH) is a graduate degree that provides training in epidemiology, biostatistics, behavioral health sciences, environmental health sciences, and health services administration.

Master of Translational Research (MTR) is a graduate program for individuals with degrees in health-related fields (MD, PhD, PharmD, DDS) that trains them with the academic and research skills needed to be competitive for independent research.

Medical College Admissions Test (MCAT) is a standardized, multiple-choice examination designed to assess a student's problem solving, critical thinking, writing skills, and knowledge of science concepts and principles prerequisite to the study of medicine. Scores are reported in Verbal Reasoning, Physical Sciences, Writing Sample, and Biological Sciences (http://www.aamc.org/students/mcat/about/start.htm).

Medical Minority Applicant Registry (Med-MAR) is a program sponsored by the AAMC to enhance admission opportunities for groups currently underrepresented in medicine. The program's registry distributes web-based basic biographical information about the examinee and the examinee's MCAT scores to minority affairs and admission offices of AAMC-member schools and certain health-related agencies interested in increasing opportunities for students who self-identify as underrepresented or economically disadvantaged individuals (http://www.aamc.org/students/minorities/resources/medmar.htm).

Medical School Admissions Requirements (MSAR) is a book published by the AAMC that includes information on medical school application procedures and deadlines; selection factors such as MCAT and GPA data; medical school class profiles, costs, and financial aid packages; MD-PhD and other combined degrees; graduates' specialty choices; and updated USMLE policies (https://services.aamc.org/Publications/index.cfm?fuseaction=Product.displayForm&prd_id=226&prv_id=276).

Medical Scientist Training Program (MSTP) is a combined MD-PhD degree training program for careers in biomedical research and academic medicine. Grants from the National Institute of General Medical Sciences, National Institutes of Health, support students in graduate training in the biomedical sciences and clinical training offered through medical schools. Graduates pursue careers in basic biomedical or clinical research.

Minority Access to Research Careers (MARC) is a program offered by the National Institute of General Medical Sciences, National Institutes of Health. It offers special research training support to 4-year colleges and universities with enrollments of minorities such as African-Americans, Hispanic-Americans, Native Americans, Alaska Natives, Hawaiian Natives, and natives of the U.S. Pacific Islands.

National Institutes of Health (NIH) is a part of the U.S. Department of Health and Human Services and supports medical research at academic centers and 27 federal institutes and centers in the United States (http://www.nih.gov/about/NIHoverview.html).

National Science Foundation (NSF) is an independent federal agency that is the funding source for approximately 20% of all federally supported basic research conducted by U.S. colleges and universities. It funds research in fields such as biology, mathematics, computer science, and the social sciences.

PhD (Doctor of Philosophy) is a graduate degree that includes both coursework and a thesis or dissertation consisting of original research worthy of publication in a peer-reviewed journal. Although the degree requirements vary considerably at different schools, the degree program in science typically takes 4 years to complete.

Post-baccalaureate programs are 1 or 2 years in length and are designed to provide academic classes with research experiences to improve a college graduate's application credentials. They may help individuals who seek a career change, increase minority participation in the health sciences, or enhance the academic or research credentials of individuals who otherwise are not competitive for advanced degree programs. Programs generally focus on preparing students for medical school, research, or the health sciences.

Problem-based learning is a teaching strategy frequently used in medical schools. It is a student-centered instructional strategy in which students collaboratively solve problems.

Scholastic Aptitude Test (SAT) is a standardized examination that is used by U.S. colleges as part of their admissions process to evaluate applicants.

Situational interviews involve asking an individual to respond to a specific situation they may have or could encounter. These types of questions are designed to draw out your analytical and problem-solving skills as well as how you handle problems with short notice and minimal preparation.

Test of English as a Foreign Language (TOEFL) is an examination that measures the ability of non-native speakers of English to use and understand English as it is spoken, written, and heard. The test is given either as an Internet-based test (iBT) or a paper-based test (PBT), depending on which format is offered at a test center.

U.S. Medical Licensing Examination (USMLE) is a series of three examinations that assess a physician's ability to apply knowledge, concepts, and principles and to demonstrate fundamental patient-centered skills that are important in health and disease and that constitute the basis of safe and effective patient care.